Bioinformatics

Bioinformatics

Colin Davenport

SYRAWOOD
PUBLISHING HOUSE

New York

Published by Syrawood Publishing House,
750 Third Avenue, 9th Floor,
New York, NY 10017, USA
www.syrawoodpublishinghouse.com

Bioinformatics
Colin Davenport

International Standard Book Number: 978-1-64740-011-8 (Hardback)

Cataloging-in-Publication Data

Bioinformatics / Colin Davenport.
    p. cm.
Includes bibliographical references and index.
ISBN 978-1-64740-011-8
1. Bioinformatics. 2. Computational biology. 3. Systems biology.
4. Biology--Data processing. 5. Information science. I. Davenport, Colin.
QH324.2 .B56 2020
570--dc23

TABLE OF CONTENTS

Permissions

Index

PREFACE

The field of bioinformatics is concerned with the development of methods and software tools used to understand biological data. It is an interdisciplinary field that encompasses techniques and concepts from biology, computer science, statistics, mathematics and information engineering for analyzing and interpreting biological data. Some of the applications of this field are identifying genes and single nucleotide polymorphisms. This helps in understanding the genetic basis of diseases, unique adaptations and differences between populations. It also tries to understand the principles of protein sequences. Some of the major areas of interest within this field are sequence analysis, gene and protein expression and cellular organization. This book elucidates the concepts and innovative models around prospective developments with respect to bioinformatics. Some of the diverse topics covered in this book address the varied aspects of this field. It is appropriate for those seeking detailed information in bioinformatics.

A detailed account of the significant topics covered in this book is provided below:

Chapter 1- Bioinformatics is a multidisciplinary field which is involved in the development of methods and software tools for understanding biological data. It deals with sequence analysis such as DNA sequencing, comparative genomics analysis, sequence assembly, etc. This is an introductory chapter which will introduce briefly all these significant aspects of bioinformatics.

Chapter 2- Computation biology is a domain that is concerned with the development and application of data-analytical and theoretical methods, computational simulation techniques and mathematical modelling, to study biological, ecological, behavioural and social systems. These applications of computational methods have been thoroughly discussed in this chapter.

Chapter 3- There are several branches of bioinformatics such as translational bioinformatics, integrative bioinformatics and structural bioinformatics. The topics elaborated in this chapter will help in gaining a better perspective about these branches of bioinformatics and their applications.

Chapter 4- Sequence analysis seeks to understand the structure, feature and function of a DNA, RNA or peptide sequence using a broad range of analytical methods. Gene prediction, DNA sequencing and whole genome sequencing are some of the fields of study within this discipline. The chapter closely examines these key concepts of gene and DNA sequence analysis to provide an extensive understanding of the subject.

Chapter 5- The methods of determining the amino acid sequence of protein or peptide is known as protein sequencing. The entire set of proteins which can be expressed by an organism is known as proteome. Its identification as well as quantification is called proteomic analysis. This chapter discusses in detail the theories and methodologies related to proteome and protein sequence analysis.

Chapter 6- There are a variety of different software which are used in the field of bioinformatics, such as EMBOSS, eProbalign and geWorkbench. The diverse applications of these bioinformatics software have been thoroughly discussed in this chapter.

It gives me an immense pleasure to thank our entire team for their efforts. Finally in the end, I would like to thank my family and colleagues who have been a great source of inspiration and support.

Colin Davenport

Chapter 1

Understanding Bioinformatics

Bioinformatics is a multidisciplinary field which is involved in the development of methods and software tools for understanding biological data. It deals with sequence analysis such as DNA sequencing, comparative genomics analysis, sequence assembly, etc. This is an introductory chapter which will introduce briefly all these significant aspects of bioinformatics.

Bioinformatics involves the integration of computers, software tools, and databases in an effort to address biological questions. Bioinformatics approaches are often used for major initiatives that generate large data sets. Two important large-scale activities that use bioinformatics are genomics and proteomics. Genomics refers to the analysis of genomes. A genome can be thought of as the complete set of DNA sequences that codes for the hereditary material that is passed on from generation to generation. These DNA sequences include all of the genes (the functional and physical unit of heredity passed from parent to offspring) and transcripts (the RNA copies that are the initial step in decoding the genetic information) included within the genome. Thus, genomics refers to the sequencing and analysis of all of these genomic entities, including genes and transcripts, in an organism. Proteomics, on the other hand, refers to the analysis of the complete set of proteins or proteome. In addition to genomics and proteomics, there are many more areas of biology where bioinformatics is being applied (i.e., metabolomics, transcriptomics). Each of these important areas in bioinformatics aims to understand complex biological systems.

Many scientists today refer to the next wave in bioinformatics as systems biology, an approach to tackle new and complex biological questions. Systems biology involves the integration of genomics, proteomics, and bioinformatics information to create a whole system view of a biological entity.

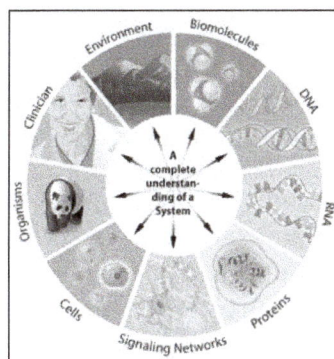

Figure: The Wheel of Biological Understanding. System biology strives to understand all aspects of an organism and its environment through the combination of a variety of scientific fields.

For instance, how a signaling pathway works in a cell can be addressed through systems biology. The genes involved in the pathway, how they interact, and how modifications change the outcomes downstream, can all be modeled using systems biology. Any system where the information can be represented digitally offers a potential application for bioinformatics. Thus bioinformatics can be applied from single cells to whole ecosystems. By understanding the complete "parts lists" in a genome, scientists are gaining a better understanding of complex biological systems. Understanding the interactions that occur between all of these parts in a genome or proteome represents the next level of complexity in the system. Through these approaches, bioinformatics has the potential to offer key insights into our understanding and modeling of how specific human diseases or healthy states manifest themselves.

The beginning of bioinformatics can be traced back to Margaret Dayhoff in 1968 and her collection of protein sequences known as the Atlas of Protein Sequence and Structure. One of the early significant experiments in bioinformatics was the application of a sequence similarity searching program to the identification of the origins of a viral gene. In this study, scientists used one of the first sequence similarity searching computer programs (called FASTP), to determine that the contents of v-sis, a cancer-causing viral sequence, were most similar to the well-characterized cellular PDGF gene. This surprising result provided important mechanistic insights for biologists working on how this viral sequence causes cancer. From this first initial application of computers to biology, the field of bioinformatics has exploded. The growth of bioinformatics is parallel to the development of DNA sequencing technology. In the same way that the development of the microscope in the late 1600's revolutionized biological sciences by allowing Anton Van Leeuwenhoek to look at cells for the first time, DNA sequencing technology has revolutionized the field of bioinformatics. The rapid growth of bioinformatics can be illustrated by the growth of DNA sequences contained in the public repository of nucleotide sequences called GenBank.

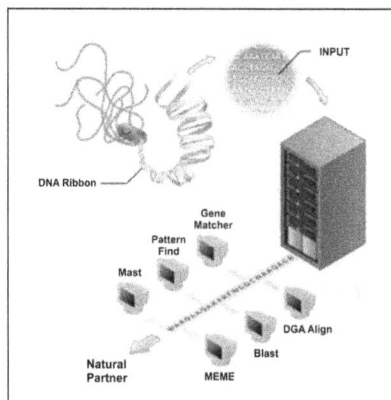

Figure: The Use of Computers to Process Biological Information. The wealth of genome sequencing information has required the design of software and the use of computers to process this information.

Genome sequencing projects have become the flagships of many bioinformatics initiatives. The human genome sequencing project is an example of a successful genome sequencing project but many other genomes have also been sequenced and are being sequenced. In fact, the first genomes to be sequenced were of viruses (i.e., the phage MS2) and bacteria, with the genome of Haemophilus influenzae Rd being the first genome of a free living organism to be deposited into the public sequence databanks. This accomplishment was received with less fanfare than the completion of the human genome but it is becoming clear that the sequencing of other genomes is an important step for bioinformatics today. However, genome sequence by itself has limited information. To interpret genomic information, comparative analysis of sequences needs to be done and an important reagent for these analyses are the publicly accessible sequence databases. Without the databases of sequences (such as GenBank), in which biologists have captured information about their sequence of interest, much of the rich information obtained from genome sequencing projects would not be available.

The same way developments in microscopy foreshadowed discoveries in cell biology, new discoveries in information technology and molecular biology are foreshadowing discoveries in bioinformatics. In fact, an important part of the field of bioinformatics is the development of new technology that enables the science of bioinformatics to proceed at a very fast pace. On the computer side, the Internet, new software developments, new algorithms, and the development of computer cluster technology has enabled bioinformatics to make great leaps in terms of the amount of data which can be efficiently analyzed. On the laboratory side, new technologies and methods such as DNA sequencing, serial analysis of gene expression (SAGE), microarrays, and new mass spectrometry chemistries have developed at an equally blistering pace enabling scientists to produce data for analyses at an incredible rate. Bioinformatics provides both the platform technologies that enable scientists to deal with the large amounts of data produced through genomics and proteomics initiatives as well as the approach to interpret these data. In many ways, bioinformatics provides the tools for applying scientific method to large-scale data and should be seen as a scientific approach for asking many new and different types of biological questions.

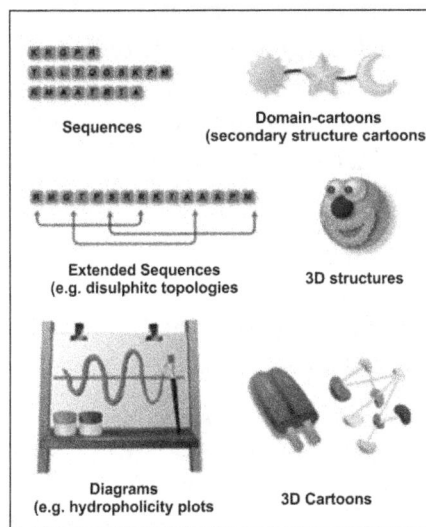

Figure: Potential Types of Bioinformatic Data. Computer based databases of biological information enables scientist to generate all sorts of data, from generating protein sequence and predicting protein domains to even producing 3D structures of proteins.

The word bioinformatics has become a very popular "buzz" word in science. Many scientists find bioinformatics exciting because it holds the potential to dive into a whole new world of uncharted territory. Bioinformatics is a new science and a new way of thinking that could potentially lead to many relevant biological discoveries. Although technology enables bioinformatics, bioinformatics is still very much about biology. Biological questions drive all bioinformatics experiments. Important biological questions can be addressed by bioinformatics and include understanding the genotype-phenotype connection for human disease, understanding structure to function relationships for proteins, and understanding biological networks. Bioinformaticians often find that the reagents necessary to answer these interesting biological questions do not exist. Thus, a large part of a bioinformatician's job is building tools and technologies as part of the process of asking the question. For many, bioinformatics is very popular because scientists can apply both their biology and computer skills to developing reagents for bioinformatics research. Many scientists are finding that bioinformatics is an exciting new territory of scientific questioning with great potential to benefit human health and society.

The future of bioinformatics is integration. For example, integration of a wide variety of data sources such as clinical and genomic data will allow us to use disease symptoms to predict genetic mutations and vice versa. The integration of GIS data, such as maps, weather systems, with crop health and genotype data, will allow us to predict successful outcomes of agriculture experiments. Another future area of research in bioinformatics is large-scale comparative genomics. For example, the development of tools that can do 10-way comparisons of genomes will push forward the discovery rate in this field of bioinformatics. Along these lines, the modeling and visualization of full networks of complex systems could be used in the future to predict how the system (or cell) reacts, to a drug, for example. A technical set of challenges faces bioinformatics and is being addressed by faster computers, technological advances in disk storage space, and increased bandwidth, but by far one of the biggest hurdles facing bioinformatics today, is the small number of researchers in the field. This is changing as bioinformatics moves to the forefront of research but this lag in expertise has lead to real gaps in the knowledge of bioinformatics in the research community. Finally, a key research question for the future of bioinformatics will be how to computationally compare complex biological observations, such as gene expression patterns and protein networks. Bioinformatics is about converting biological observations to a model that a computer will understand. This is a very challenging task since biology can be very complex. This problem of how to digitize phenotypic data such as behavior, electrocardiograms, and crop health into a computer readable form offers exciting challenges for future bioinformaticians.

Ontologies for Bioinformatics

Ontologies are a concept imported from computing science to describe different conceptual frameworks that guide the collection, organization and publication of biological data. An ontology is similar to a paradigm but has very strict implications for formatting and meaning in a computational context. The use of ontologies is a means of communicating and resolving semantic and organizational differences between biological databases in order to enhance their integration. The purpose of interoperability (or sharing between divergent storage and semantic protocols) is to allow scientists from around the world to share and communicate with each other.

During the 1960s, there was a simultaneous evolution of digital protein and taxonomic inventories. By the 1980s, these had matured and were institutionalized with an attendant proliferation of biological data. These datasets were, however, maintained in closely-guarded proprietary repositories or 'silos' with little or no communication between them. The 1990s were marked by a shift in emphasis from accumulating vast volumes of data to reducing overlap between databases and making use of extant data across various repository locations. This process of increasing communication between databases is known as interoperability—the focus of which is to enable data sharing and comparison.

As the cumulative body of biological knowledge increases, generating a comprehensive and consistent account of biology hinges upon the ability of scientists to draw upon and synthesize vast datasets across distributed digital resources. The ultimate objective of biodiversity informatics is to generate a "global inventory of [all] life on Earth", and is premised on the seamless digital accumulation of distributed taxonomies. Because contemporary biological—particularly 'omics' and model organism—databases stress data at the molecular scale, they do not adequately represent the physiology they describe. There is thus a need to compile the cellular features of those organisms into discernible representations of those organisms themselves.

The rise of 'omics' science—genomics, proteomics, and metabolomics for the identification and prediction of genetic product components, signatures, and processes has contributed the molecular-level information upon which a systems view of biology is predicated. Certainly the complexity of biology resides at the level of gene products. In this way biodiversity can be understood as the compendium of the biology of organisms. In a computational environment, biodiversity is the hereditary information encapsulated within genetic products and identified via the collective of mappings of several model organism genomes. While the maturation of 'omics' has been facilitated in large part by the capacity to seamlessly make divergent data sources interoperable, it has presented a new set of engineering challenges. These include the need to integrate diverse and remote data sources as well as to extract knowledge from digital information post-integration.

The paradigm shift the 'omics' revolution has created within biology is best exemplified by gene prediction (also known as gene finding), and functional prediction tasks. New technologies such as micro arrays generate huge and ever-changing volumes of data. The rapid growth of genome mapping necessitates the ability to automate gene-calling, or the identification of the individual genes of a genome. Gene finding involves algorithms for the identification of biologically functional regions—or exons—of sequences which explicitly code for proteins. These are referred to as coding regions. The objective of automated gene prediction is thus to determine the "coding potential" of genetic sequences. This process uses self-learning algorithms which predict unique signatures of the genetic spectrum that indicate distinct clusters of material. Where genes have been located, the biological functions of many protein sequences are as yet undetermined. Gene finding and functional prediction go hand in hand and are rarely treated separately as researchers often desire to discern the roles of newly identified gene products. The potential for predicting protein function similarly rests on its inference over incompletely annotated sequences on the basis of homologues in other species. However, neither is an easy feat as the coding regions of eukaryotic organisms are both sparse and small, making the identification of exon/intron boundaries—and thereby the identification of protein function—difficult, resulting in erroneous gene annotation. In the present era of functional genomics, knowledge production is however dependent on the ability to recover genes and proteins on the basis of their (correctly) annotated functionality, pathways, and/or protein-protein interactions. This is no trivial task; indeed it necessitates the resolution of semantics, or differences in meaning and naming conventions between distributed data resources.

Unlike systems architectures, the integration of which constitutes an 'IT problem', data are not semantically transparent. Although a structural linkage can now be easily defined between data sources such that a user can retrieve data on the basis of standardized queries across data sources with conflicting database schemas, this does not render the results of those queries meaningful. A prime example is the notion of 'gene'—the primitive of modern biology. While the concept of 'gene' is still evolving, two dominant concepts exist: the Human Genome Database defines a gene as a DNA fragment that can be interpreted as (analogous to) a protein; whereas GenBank and the Genome Sequence Database (GSDB) consider a gene to be a "region of biological interest with a name and that carries a genetic trait". Two databases can be developed based on different understandings of 'gene'. As a result, retrieving data from semantically orthogonal databases on the basis of a 'gene' keyword search can initiate error propagation—in this case in the form of false analogues—in the analysis and subsequent results. The complexity of biological terms exacerbates this problem. Even where two variables in disparate databases are semantically equivalent, their relations to other knowledge objects in the data repository may not be. This is referred to as schematic incompatibility and refers to the relative position of the term in a taxonomic hierarchy.

In order to accommodate both semantic and schematic differences between biological databases, 'omics' research requires a method of expressing the contexts from which biological concepts emerge—at the database level. This is because functional prediction hinges upon the identification not just of sequence homologues but similar cellular components participating in a similar biological process. The component cellular, molecular, and biological details are often located in separate data sources, a function of the narrow scope of biological information produced by any given laboratory. Exploiting the vast digital resources of biological data for prediction services requires that the cellular, molecular, and biological contexts of proteins be adequately encoded and furthermore machine-readable.

Ontologies—or the use of a singular taxonomic and knowledge representation schema—are a way of resolving these semantic issues between databases. The bioinformatics literature has been heavily promoting ontologies as an operational solution for biological interoperability since the turn of the millennium. Much of this literature assumes that the reader has a prior understanding of computing and is delivered in impenetrable technical language or emphasizes a singular aspect of ontologies in biology.

The power of ontologies lies in their capacity to provide context for biological semantics.

Ontology has traditionally been understood to be the essence of being—or what something really is. In the information sciences, an ontology is a fixed universe of discourse in which each element (e.g. field name or column in a database) is precisely defined. In addition, each possible relationship between data elements is parametized or constrained. For example, DNA may comprise chromosomes but not the reverse. In an ontology, these relationships are made explicit formally.

The prefix 'formal' refers to the property of machine-readability. In other words, a *formal ontology* is a machine-readable model of the objects allowed into a formal universe and their associations or relationships between them upon which some automated reasoning tasks can be performed. In a formal environment, an ontology constitutes a surrogate of knowledge abstracted from the real world—in this case, the cumulative body of biological science—in a coded form that can be translated into a programming language.

Scientific or systems ontologies contain three levels of formalization. The first is the conceptual, which is then translated into a formal model of the data elements in the ontology (e.g. proteins) and the possible relationships between them. The final stage or level is the development of code that can be run by computers. Ontologies are structured much like a biological taxonomy with general concepts appearing at the top of the tree and becoming more general as one traverses down. The hierarchical schema, however, is only a 'shell' that can accommodate the concepts and their relations particular to a domain. It must be populated by domain knowledge expressed in a formal semantics—a computing syntax such as a markup language—that allows all entities declared into the ontology to be precisely defined and their interrelationships given strict parameters with the goal of enabling realistic biological models.

Formal semantics permit the distinction of concepts declared into the model. To satisfy the strict criteria of formal ontology building, the formal semantics used to instantiate an ontology should be premised on a formal logics particular to some logical algebra —such as description logics (DL)— which contain predetermined rules for "when two concepts are the same, when one is a kind another, or how they differ". These rules must furthermore be expressed in some machine-readable syntax—in this case, a knowledge representation language such as the Web Ontology Language (OWL). Such rules govern the expression and processing of *relations* between concepts in the hierarchy. Relational expressions are the implementation basis for all subsequent computing and modeling tasks in a software environment. illustrates the progress from concept to code. Formalization is the basis for the transition from a conceptual entity to a machine-readableform.

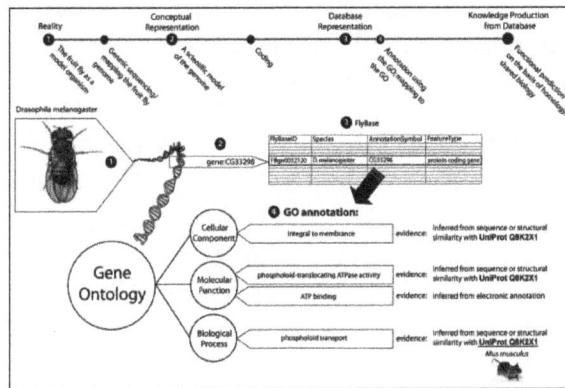

The formalization process. Moving from a concept of a particular gene to its encoded reification and ontological representation. Note how the entity (fruit fly) becomes increasingly represented in digital database format as it is formalized, or abstracted from its real-world form. The entity loses dimensionality, while researchers gain the advantage of computational function. Figure illustrates in more detail the role that entity descriptions—or annotations—play in creating a larger standardized digital knowledge environment for bioinformatics. Note that any gene product many have more than one annotation in the same branch, and can be annotated in three different branches of GO (Cellular Component, Biological Process, and Molecular Function).

The ability to define relationships between concepts distinguishes formal ontologies from earlier integration and interoperability approaches. How they are expressed are detailed in the subsequent sections on the GO and OBO efforts. Relationships are an expression of the *context*—akin to usage in natural language—in which concepts are used or from which they emerge. The utility of capturing relationships between concepts is thus that they convey semantics; content semantics are expressed by identifying how concepts relate to each other in the hierarchical knowledge space. This hierarchical knowledge space is a parent-child structure that conveys the semantic granularity of the relation between any two concepts by rendering entities to be either more specific of more general than each other. This formal ontological structure implies at least one kind of relation: a hyponymic (*is-a*) relationship is implied by the hierarchical nesting of terms and denoted by their position relative to each other in the family tree on the basis of their subsumption (where a concept is a subclass or member of another) and specialization (where one concept is the superclass of or contains another). Additional relationships can be asserted between concepts as a directional association (i.e. the relationship proceeds from one concept to another). Relationships—referred to as *properties*—are akin to 'semantic edges' which depict the meaning of data elements by providing the *context* of their usage (where context is analogous to how concepts participate in class membership).

Formal ontological expressions are stated as propositional triplets consisting of *concepts* (real-world entities that populate the model), their *properties* (or relationships between said entities), and *instances*(particular occurrences of a concept; for example, a particular gene with its own unique identifier in a database) in a hierarchical model. A triplet (concept + property + instance) constitutes a proposition, or "definitive statement about (part of) the world". Where an ontology is formal in the sense that it is underwritten on an axiomatic logic such as Desription Logics (DL), the axioms of the logic can be applied to impose *restrictions* that define conditions under which concepts in a domain logically participate in relationships with each other. For example, we can impose a cardinality restriction to specify that, following the series of generic examples provided by, a "man" must have at least one testes.

In a strict definition of formal ontology, the axiomatic logics serve to underwrite a formal notation for content specification. For example, DL comprise the logical *semantics* for knowledge representation which constitute the basis of ontological encoding specifically designed for a group of knowledge representation languages which include OWL, the standard language for ontologies over the Web. The eXtensible Markup Language provides the tag-based syntax for OWL, whereas its schema is defined by the Resource Description Framework (RDF), which specifies what the 'triplet' structure (*concepts* + *properties* + *instances* described above) of ontological expression. A standard schema ensures that when OWL statements are *parsed* or transformed into the component data structure of the target formal ontology, the parser knows which part of the expression constitutes the concept, which section the relation, and which the instance. It is this structure which makes the grammar of an ontology *meaningful*—in the case of bioinformatics, for example, it anchors annotations to the gene products they characterize.

This structure moreover makes the ontological model amenable to implementation in a software environment in order to allow for the kind of intelligence described using the example of a cardinality restriction in the instance, 'a man must have at least one tests'. The taxonomic structure of formal ontologies captured using logical notation and expressed in a knowledge representation language allows the semantics of concepts to be computed on the basis of concept inheritance. This is known as *reasoning*, where an application infers non-explicit (not directly stated) relationships between concepts. For example, where two proteins identified using different unique identifiers in disparate databases are described as participating in the same biological function, being part of the same sequence, having the same cellular location, etc., they can be recognized as referring to the same concept and can thus be extracted from separate databases on the basis of these functional characteristics rather than nominal IDs. The ability for each term to relate to every other term in the hierarchy is a way of capturing—and expressing—the complexity of biology.

Reasoning can therefore be thought of as supporting both inference and query. Inference consists of computing the hierarchy—for example, it will reveal multiple inheritance amongst classes as mentioned aboved. Query consists of the ability to interrogate the concept hierarchy on the basis of object associations or, conversely, to reveal object associations amongst selected concepts or classes. Imposing the above cardinality restriction therefore has two implications. The first is that any data object labeled or identified "man" in a data repository mapped to an ontology with the above property restriction imposed upon the man-testes relation will be recognized as a (likely person) with at least one testes. Conversely, the execution of reasoning tasks on the ontology or any data structure mapped to it will compute whether all instances of man are consistent with (a person) having at least one testes.

Ontologies—with their hierarchical structures—capture the semantic granularity of biological databases. The property of inheritance allows the computer to process, for example, that the concepts used to annotate two respective sequences are both 'children' of the same meta-concepts (i.e. they are a kind or part of a the same overarching concept; alternatively, members of the came class). This permits researchers to locate regions of exact correspondence as well as those with a high degree of similarity. Entities may relate but are not synonymous—for example, where 'protein' is a subclass of another concept, 'gene products'. This does not dictate that proteins and genetic products are one and the same, but rather allows the expression of a membership relation at a much finer semantic resolution such that proteins can be understood as one, but not the sole, kind of gene product (which also includes RNA).

The problem of semantic and schematic heterogeneity and introduced ontologies as a means of mitigating the problem.

GO: Ontology in Practice

The use of ontologies for bioinformatics is being driven by the proliferation of genome-scale data-sets and the diffusion of the Internet and its protocols for data sharing and exchange. Bio-ontologies fulfill two central functions for the biological domain—first, they "clarify scientific discussions" by providing the vocabulary and terms under—and with which—such discussions take place, and second, they enable data discovery across distributed data resources. The pre-eminent bio-ontology is the (GO), a Web-based, open source knowledge resource for bioinformatics and the second-most cited biological data resource after UniProt.

The GO project evolved as a joint endeavor between three model organism databases: FlyBase, Mouse Genome Informatics Database (MGI), and the Saccharymyces (yeast) Genome Database in 1999. The formation of the Gene Ontology Consortium (GOC) coincided with the successful completion of the mappings of several eukaryotic genomes. The key to associating these model databases was the genetic structure of organisms. A potential problem lay in that these databases had been designed and populated with competing concepts for gene. Moreover, there was still limited understanding as to how the located genes were controlled and more importantly what functions many of these served. As there is a high degree of functional conservation in homologous organisms, gene function can be reasonably inferred through probable genetic orthologues. In other words, rather than 'reinventing the wheel', biologists and bioinformaticians could transfer functional attributes describing the cellular behaviors of gene products between these databases thereby significantly reducing workload.

The chief impediment to this task were not the unique identifiers for the gene products themselves as researchers had been tapping into protein and gene databases such as GenBank and Swiss-Prot, TrEMBL and PIR for decades (the latter three joined to form the Universal Protein or UniProt protein repository in 2002). Because sequences are unique, they could be easily accessed on the basis of sequence characteristics (though there was sequence redundancy between protein repositories). Computationally, because sequences can be quantified, this is a trivial integration task that simply requires the normalization of unique codes. Rather, it was the functional descriptions of gene products that proved challenging. Integration had to proceed within the context of the molecular and biological characteristics of each gene product identified. In an attempt to solve the problem, informatics experts from the three original participating model organism databases devised functional classification systems in the hopes that these precursors to the GO would facilitate interoperability. What soon became apparent, however, was that these functional classifications were not common between organisms.

In other words, the annotation was not consistent from one database to the next. Gene annotation is defined as the "task of adding layers of analysis and interpretation to raw sequences". This includes information about their function, position relative to coding/non-coding boundaries, participating process, etc. Annotations constitute a set of metadata, or 'data about data' Historically, annotation has been stored as free-text or at best semi-structured descriptions semantically particular to the terminological or classification systems unique to many of the databases. There were two challenges. First the use of competing nomenclatures precluded the linear association of database semantics. Second, the expression of these annotations in natural language provided little context for data mining because they were not machine-readable. Returning to the example of functional prediction, protein functions are inherently dependent upon context, particularly cellular context. This is exacerbated in the case of proteins particularly as many sequences often have multiple functions.

The GO Consortium formed as a response to the pervasive semantic heterogeneity of biomedical data and its lack of formality. Indeed the GO was designed for making historically free-text based annotations tractable. The three participating database programs agreed to work in concert to provide the biological community with a consensus-driven framework to guide the annotation of gene products such that their structure (e.g. how molecular function is described and which part of the description occurs in what syntactic order) and semantics (the terms and concepts) are consistent. The result was the GO—a "structured, precisely defined, common, controlled vocabulary for describing the roles of genes and gene products in any organism". The GO is not a taxonomy or index of all known proteins and gene products, but rather provides a standardized set of names for genes and proteins and the terms for characterizing—or 'annotating'—their behaviors.

Gene product semantics are organized into three categories which capture the primary 'aspects' of genes: i) biological process, which captures the larger process in which the gene product is active; ii) molecular function, the biochemical function a gene product contributes to that process, and iii) cellular component, the location in the cell where that particular function is fulfilled or expressed. Concepts or terms constitute nodes, and vectors referred to as edges represent relationships between concepts. These three sub-ontologies are maintained independently because the one-to-many relationships between process, function and cellular location would make a singular graph representation intractable. Annotations for the same term in each 'view' are cross-referenced on the basis of a unique identifier or serial number assigned to each term in the GO. Increasingly, these identifiers are being used to refer to concepts in other protein and gene-oriented databases and constitute a linear and direct means of mapping databases to the GO. A 2005 figure estimates the GO as consisting of more than 17, 500 terms distributed amongst the three subgraphs. All possible annotations for a protein can be represented using these concepts.

Each of these three separate annotation categories—biological process, molecular function, and cellular component—is represented as its own directed acyclic graph, or DAG. A DAG is a data structure similar to a tree which represents knowledge hierarchically, mirroring the taxonomic structure of biological knowledge. Any entity can point to any other entity in the mathematical space; this is, however, a direction, and non-recursive, encoding. In other words, concepts can point to other entities in the model, but those entities do not 'point back' as in OWL. Indeed the DAG can be considered to be the native knowledge representation (KR) language of the GO. Unlike the KR languages introduced above, however, DAG semantics are not predicated on a formal logic as they are in the case of OWL. Rather, machine readability is instructed by the directional links between pairs of concepts. Semantic 'edges' (relationships) in the DAG are simply "ordered pairs of nodes". Pointers are like edges in the sense that their semantics are directed, and are labeled with the relationship that associates related classes. These associations are of only two relations: is-a, which denotes that concepts are kinds of entities, and part-of, which can signify the participation or contribution of a concept in a sequence or process.

The DAG is available in many file formats—XML, OWL—but the most common formal notation in which GO ontologies are rendered is the Open Biomedical Ontology flat file structure which is underwritten by a modified subset of Web Ontology Language (OWL) description logics (DL) concepts for content specification.

Like OWL, OBO is an ontology language, and standard 'file format' for GO annotations. It is however less expressive than OWL. These relations are unidirectional and linear as per the DAG data model and do not require the recursive relational declarations (where the reciprocal or inverse of a relationship is also encoded) characteristic of OWL statements. Thus a flat file structure that only supports sequential reading is appropriate for the GO because relations are read from broader or general to more specific or precise concepts.

At the level of the database, the GO is represented as a structured vocabulary; more specifically, as gene product annotations expressed using concepts and their tripartite (biological, molecular, and cellular) structure defined in the GO. The GO is not considered an informatics ontology in the full sense of the term because it has not been designed to be deployed within software environments which execute semantic inference on the basis of logical semantics. Moreover it does not fulfill the conditions of formality identified by. Rather it is considered and referred to by its engineers as a "controlled vocabulary" The nevertheless has many of the characteristics of a formal ontology: machine-readability, formal notation, a hierarchical knowledge structure, and relational associations between concepts. In other words, the GO may be considered a partial implementation that uses many concepts of formal ontology. Part of the reason, however, that the GO is only a partial implementation is that it was designed to be operational within existing infrastructures, requiring no changes to existing architectures.

Not with standing, the GO provides the standard vocabulary for semantic integration and automated tasks for bioinformatics. As such it is more than merely a sophisticated data dictionary.

Whereas controlled vocabularies or data dictionaries provide a definition of the terms used by a community of practice and these may indeed be machine-readable and thereby formal, a nomenclature does not capture the hierarchical representation of knowledge nor the corresponding relations between all concepts in the data space, and thereby does not support computational reasoning. Several terminological systems such as SNOMED (Systematized Nomenclature for Medicine) and MeSH (Medical Subject Headings) have, however, been mapped to the GO.

The GO is a global ontology. In other words, it is a central knowledge proxy to which other ontologies or knowledge representations may be aligned. Ontology mapping is the process of defining associations between ontologies. This involves the formal declaration of relational links between entities, much like that involved in relating concepts in a hierarchical ontological structure. Ontologies can either be aligned whereby the formalisms remain separate entities but are related, or merged wherein a singular ontology is generated from the crossproducts of two input ontologies. 'Mapping' is thus unidirectional and always from the constituent database to the GO. illustrates the role that GO plays in development of global biological ontology and the mechanics involved.

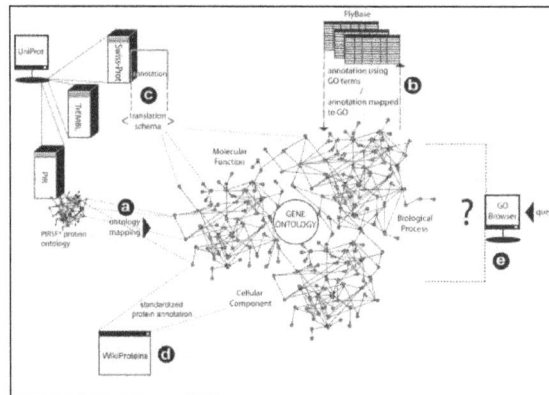

The Gene Ontology as a global ontology for bioinformatics. Smaller scale bioinformatics ontologies almost invariably map to the GO (a). Several large databases, such as FlyBase (b), contribute annotation to the GO using its semantics such that there is a direct mapping between genes/gene products at the database level and their participation in the ontology. (FlyBase annotation is explained in greater detail in Figure). Where annotation is unique to the database, a translation program can transform annotation into a tractable GO representation (c). The GO provides a standardized vocabulary for the description of genes and gene product across not only databases but also in emerging bioinformatics infrastructures, such as WikiProteins (d). The consistency of semantics reduces ambiguity in the query of bioinformatics resources, and allows genes and gene products to be retrieved on the basis of common biology rather than lexical coincidence (e).

A global ontology paradigm is appropriate for the domain because there is a finite, though as yet not fully discovered or known, body of genetic information shared between all life on Earth. Accordingly there is no need to build *local* ontologies each capturing a competing account or version of a biological universe. Such a scenario would be much more intensive, requiring the definition of linkages between each participating ontology. A global ontology serves as a *proxy context* which interfaces all participating knowledge formalisms are translated to the unique semantic points of the proxy and then compared on the basis of this translation.

The alignment of currently non-compatible ontologies to the GO is one avenue for its curation or the process of developing and contributing content or adding value to digital knowledge representation systems such as databases or ontologies. For the GO to serve as a comprehensive knowledge resource for the biological community, it must reflect the continuously increasing body of biological, specifically genetic-level, knowledge. In other words, it must expand to keep pace with the identification of new genes, sequences, functional determinations, etc. Rather than being the responsibility of the Consortium, GO curation has been user driven from inception GO expansion efforts are supported by the scientific publication process, with several leading periodicals and sequencing initiatives mandating that newly identified sequences be deposited into GO-compliant databases and any new annotations be added to the GO. Early curation was characteristically on a need-be basis with concepts added to the GO when authors were annotating genes, etc. Such a*d hoc*practices, however, resulted in logical problems in the DAG and indeed soon became inefficient as the scope and scale of the GO has steadily grown. Increasingly, methods for contributing annotations to the GO are based on the automatic generation of annotation concept definitions on the basis of cross-products between databases (as local ontologies) and the GO itself.

The GO was designed specifically to account for molecular function, biological process, and cellular components of gene products. It lacks the semantics to describe the physical attributes of genes, to describe a protein family, or to account for experimental processes and diagnostic procedures. There are both proprietary and open ontologies with richer semantics for more specific description tasks for biology either being developed or presently available The majority, however, are designed with mapping to the GO in mind.

Ontologies in Support of Bioinformatics

The largest public contributor of annotations to the GO project is the Gene Ontology Annotation Database (GOA). While annotation is the central organizing principle and raison d'etre of the GO, the potential of their ontological encoding is not to have a hierarchically structured record of concepts used to annotate the data of biology, but rather to exploit the ontology for a series of bioinformatics services which remove the burden of data-intensive tasks from molecular biologists and moreover produce knowledge over and above facilitating its reuse.

One of the primary objectives for bioinformatics to realize is the automation of annotating cross-matches between databases. The electronic generation of annotations based on homology is particularly desirable as the manual curation of gene-oriented databases is time consuming and non-trivial for humans. The GO facilitates the automatic annotation of gene products at the database level. GOA for instance uses GO terms to generate annotations for the UniProt Knowledgebase (The consortium of SwissProt, TrEMBL, and PIR-PSD protein databases). Existing data held in UniProt are electronically associated with or translated into GO terms on the basis of a defined mapping file used to facilitate the conversion of keywords in the constituent databases to tractable GO representations. Once the semantics are consistent between data sources, biologists who have identified a new sequence, for example, can navigate the GO via an interface known as an *ontology browser* on the basis of these common data elements and indeed use the existing GO annotations to not only discover sequence similarity but to also automatically populate or their own database using the existing annotations for homologues from other curated data sources. Thus the ontology functions as a 'translation schema' This is possible because GO is underwritten by a structured grammatical framework (e.g. RDF) that predetermines the occurrence or sequence of description types in a proposition, allows the expression to be parsed and correctly broken-down such that it can be stored according to the structure of the target database.

The GO can be used to automate the following services: database annotation, GO extension (automating the transfer of new annotation concepts *to* the GO), prediction services, and database population. Prediction services supported by ontologies yield new biological knowledge. Gene location using current generation algorithms uses data from a pair of genomes to locate areas of genetic affinity; these areas of 'overlap' are often the sites of new genes. The success of this is based on the semantic consistency of annotations for the input genomes. In addition, the GO supports nuanced data exploration and query. The hierarchical structure of knowledge afforded by ontology allows the isolation of the appropriate concept for query on the basis of its context or position relative to other entities in the data space. This allows users to formulate searches using conventional keywords, but resolves the meanings of those keywords.

Once the protein or gene of interest is isolated, its location confers more information than a binary indication of its absence or presence in a database. Not only do we know about the occurrence of a protein, for example, but we are told something *about* it. The proprietary EnsemblGO Browser is an interface which compiles annotations to generate reports or summaries centered on the biological entities isolated in the GO such that, for instance, "the previously unconnected classes Antigen, Immunogen, and Adjuvant are now recognized as being objects (for example, Proteins), which participate in a certain role (as Immunogens) in a specific process (such as Immunization)".

Ontologies can further be used as a basis for exploring datasets. We have devised a methodology called ontology-based metadata which uses ontologies as a component in a metadata-based framework for the comparison of a series of eight 'near' but non-equivalent terms that have been identified as an obstacle to integrating perinatal (pregnancy and antepartum) health data registries across Canada. Our objective is to provide health researchers and data stewards with a basis for drawing meaningful parallels between data elements to enable the legitimate integration of peri-natal data registries. Ontology-based metadata for each term is first collected via a series of electronic forms which standardize the description of each concept. Each constituent database is responsible for detailing how these terms are used in their particular jurisdiction—or context. This includes a specification of the classification standard used (e.g. ICD-10), the identification of thresholds for measurement specifications, and space for free-text descriptions of any policy constraints which may influence how the term is used in a given jurisdiction. In addition to these 'annotations', we encode each perinatal database as a formal ontology in OWL. These ontologies capture the semantic structure of database terms. These are then merged into a single ontology, with the relationships between each and every concept defined in the product tree. Both the ontology-based metadata and the ontologies are inputs to a semantic data discovery portal where researchers specify which terms in two respective databases are to be compared via a graphical user interface (GUI). The application returns to the user both the encoded relationship between the concepts extracted from the OWL code—for example, where pregnancy-induced hypertension is a KIND-OF hypertension complicating pregnancy—and the ontology-based metadata for each term in the selected databases. Thus the researcher is provided with both a marker for the granularity of the semantic relationship between two concepts, as well as valuable metadata which are used to inform perinatal database decisions.

The gestational hypertension/hypertension example above would indicate that hypertension experienced during pregnancy is a more general concept which includes gestational hypertension but also encompass pre-existing hypertension. In some databases, hypertension and pregnancy-induced or gestational hypertension are not differentiated from chronic or pre-existing incidences of disease. Alternatively, in other databases, these concepts are distinguished from each other on the basis of the periodicity of disease onset such that chronic hypertension and pregnancy-induced or gestational hypertension are disjoint (database A). In yet other registries, any form of hypertension presenting during pregnancy is considered gestational such that a pre-existing condition which first manifests itself during pregnancy is still encoded as pregnancy-related (database B). There is thus a semantic incommensurability between what 'gestational hypertension' represents in databases A and B, precluding a direct mapping between these concepts indicating semantic equivalence. Rather, 'gestational hypertension' in database A would be a kind of gestational hypertension as the concept is reified in database B. If a researcher were to query 'gestational hypertension' across both databases, she would logically accept them as referring to the same concept on the basis of lexical coincidence. However, the lack of an encoded equivalence between these two concepts would preclude their conflation. Thus our this approach not only provides information regarding how concepts should be associated, but also uses formal ontologies to restrict which concepts may be legitimately compared. This nesting of relationships between semantic terms is described.

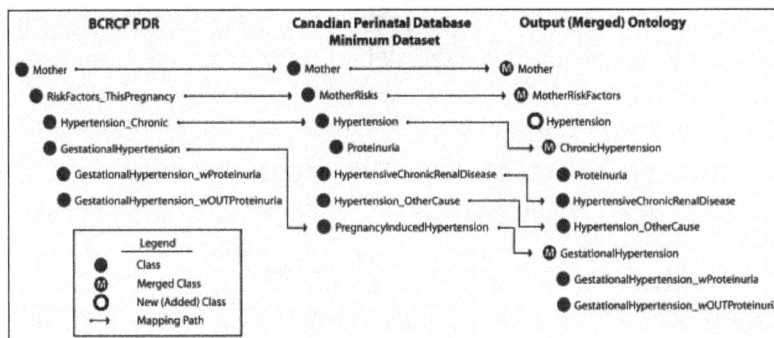

Ontology mapping. An ontology for hypertension resulting from the merging of hypertension concepts in the British Columbia Reproductive Care Program Perinatal Database Registry (BCRCP PRD) and the Canadian Perinatal Database Minimum Dataset. The resulting output ontology shows the hirerarchical nesting of hypertension semantics originating in respective databases in relation to each other.

Another instantiation of the ontology-based metadata concept similar to our implementation is WikiProteins, a structured semantic space for capturing the context—biological, physiological, chemical, etc.—of proteins and then sharing that collaborative knowledge with other biologists in real-time Historically, the problem with metadata has been that it is so labor intensive and never updated WikiProteins provides a mechanism for sharing the labor and ongoing maintenance by participants. This collaborative Web-based workspace facilitates the open curation of protein-specific information by providing biologists and bioinformaticians with a means of contributing to the cumulative body of biological knowledge. At the moment, it serves UniProt and GO descriptions for the annotation of proteins via a series of standardized forms or 'slots' for their description. This consists of definitions, attribute-value relations (e.g. a protein can be given the attribute "tissue" with the value "expressed in muscle fibers") and provisions for disambiguating sequences or instances of proteins by identifying synonyms, disjoint concepts, alternate spellings, etc. Curators can link their descriptions or proteins to other citations, references, and publications indexed in PubMed. Moreover, the wiki concept ensures that these annotations are self-validating. Other users can go in and add or revise the annotations. For example, using the "tissue" example above, a subsequent curator can reify this protein as "expressed in muscle fibers and *the brain*".

Similar to our ontology-based metadata approach, it combines both free-text fields for open description and more restrictive means of disambiguating proteins and protein concepts. For instance, it extends the ability to identify whether these synonyms are instances of equivalent meaning, or if they are different. If the latter is the case, curators can further annotate—or describe—specifically where these differences lie. WikiProteins is but one example of where the GO is being deployed to provide a standardized vocabulary for annotation across distributed data resources.

As our non-automated method for data discovery and WikiProteins for protein knowledge exchange illustrate, ontologies are not standalone solutions for interoperability but rather comprise a component of or input to large-scale interoperability infrastructures. Indeed ontologies are knowledge representations and *not* software applications, having no innate functionality. As such they must be deployed within digital architectures where constituent programs can exploit the hierarchical structure of formal ontologies to facilitate data sharing at the level of semantics. Many such cyber-infrastructures exist for biology and biomedicine. A notable example is the Cancer Biomedical Informatics Grid (caBIG), a Web-based National Institutes of Health (NIH) data consortium for cancer research. caBIG is built on an open grid architecture similar to a federated database environment where users are presented with a central interface which seamlessly integrates participating databases, but with the addition of Web services that provide tools and applications. The emphasis of caBIG is on the provision of services—such as data analysis tools, applications, scripts, algorithms, etc.—relevant to cancer research. The grid is organized into a series of "workspaces" or virtual communities where participants can both access, revise, and upload new technologies to that specific sub-domain of application or interest. The emphasis of caBIG is on services, with participants notifying each other of the constituent services they make available by means of UML (Unified Modeling Language) metadata wherein the services are described using standardized DL-annotated concepts from a vocabulary service which defines terms and concepts in biomedical vocabularies. Here, ontologies are utilized as a standardized set of concepts and terms across applications and services for their uniform description such that researchers can locate and access the appropriate technologies on a need-be basis. This provides interoperability across distributed cancer research centers at the level of services.

Sequence Analysis

Sequence Analysis is considered a standard way to analyze next generation sequencing (NGS) data for finding a genetic cause for Mendelian disorders. Standard sequence analysis enables detection of aberrant nucleotide order, most commonly single nucleotide variation (e.g. point mutations) causing either synonymous or non-synonymous change at amino acid level. Non-synonymous changes can be either missense variants (amino acid substitution by another) or nonsense variants (generation of premature stop codons). Standard sequence analysis also detects small insertions, deletions and their combinations, indels, up to 50 base pairs.

Sequence Alignment and Analysis

In general, the more similar two sequences are, the more similar should their functions be and more phylogenetically close they should be. The sequences for the same gene in a group of species will be more different the more distant phylogenetically they are. Sequences will get mutations over time, so the more time has passed since two species split, the more mutations will have their sequences and the more different their sequences will be. Mutations can be residue (nucleotide or aminoacid) substitutions, insertions or deletions.

Biological Science

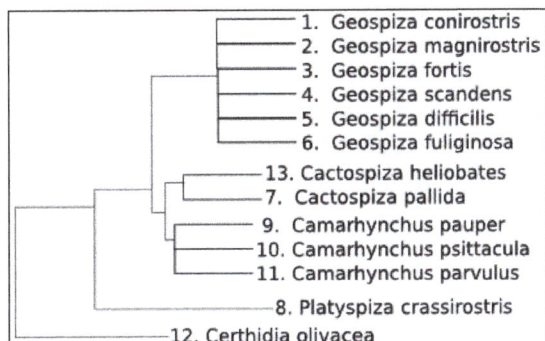

Biological sequences are similar, usually, because they are homologous, because they share a common ancestor. Homology is not a quantitative concept, two sequences can be homologous or not, but they cannot be 50% homologous. They either share a common ancestor or they don't.

How do we know that two sequences are homologous? Usually we infer it from their similarity. If two biological sequences are similar we tend to infer that they are similar because they are homologous.

Usefulness

Alignments could be used to:

- Quantify the phylogenetic distance between two sequences.

- Look for functional domains.

- Compare a mRNA with its genomic region.

- Identify polymorphisms and mutations between sequences.

Sequence Alignment

The first step to compare two sequences is, usually, to align them.

```
No alignment

 CGATGCTAGCGTATCGTAGTCTATCGTAC

                   |      ||

 ACGATGCTAGCGTTTCGTATCATCGTA

Alignned

 -CGATGCTAGCGTATCGTAGTCTATCGTAC

  ||||||||||| ||||||||||||||||

 ACGATGCTAGCGTTTCGTA-TC-ATCGTA-
```

There could be substitutions, changes of one residue with another, or gaps. Gaps are missing residues and could be due to a deletion in one sequence or an insertion in the other sequence.

Gaps complicate the alignments. Algorithms should take into account the possibility of introducing gaps and once we allow them to create gaps several alignments can be constructed between two sequences.

```
No gaps (10 matches)

 a:  ATATTGCTACGTATATCAT

          ||||||||||

 b: ATATATGCTACGTATCAT

With one gap (14 matches)

 a:  ATAT-TGCTACGTATATCAT

      |||| ||||||||||

 b: ATATATGCTACGTATCAT

With two gaps (16 matches)

 a:  ATAT-TGCTACGTATATCAT

      |||| ||||||| ||||||

 b: ATATATGCTACG--TATCAT
```

The objective of a sequence alignment is, usually, to align the homologous positions of the two sequences. The homologous positions are the ones that come from the same position in the ancestral sequence. We don't know the ancestral sequence, so we won't be completely sure that we have succeeded. Another complementary objective could be to align protein regions that have the same structure or function.

Aligning similar sequences by any algorithm usually creates alignments that are usually correct, but when sequences are very different aligning them could be a challenge. Once a long time has passed since the split of the species the sequences can be so changed by the mutations that any meaningful similarities could have been lost an creating a meaningful alignment could be very difficult.

Evaluating the Alignments

To be able to compare the different possible alignments we can score them. We can create a scoring system that gives more points to alignments that are biologically more reasonable. Ideally we would create a scoring system that gives more points to the alignments that align the homologous positions.

A naive scoring system could be to count the number of matching positions, or the number of matching positions along 100 residues. Usually the scoring systems also take into account the number of gaps. They penalize the alignments depending on the number and the length of the gaps present. So the main features taken into account to create an scoring system are usually:

- Number of matching residues (taking into account the similarity if they are aminoacids).

- Number of missmatches.

- Number of gaps.

- Length of the gaps.

We can devise different scoring schemes with those measures. For instances:

- Scoring schema 1: match +1, mismatch: 0, gap creation: -1 gap extension: -1.

- Scoring schema 1: match +1, mismatch: -1, gap creation: -1 gap extension: 0.

Of course, one alignment will have a different score under different scoring schemes. Speaking of the score of an alignment it is meaningless if we do not take into account the scoring schema used. It also has no sense to compare the scores of different alignments done under different schemes.

Once we have decided which scoring schema to use, the alignment algorithm should try to create the alignment that obtains the maximum score under that particular scoring schema.

Every software implementation of an algorithm will usually have some default values for its parameters. These default values have been calibrated by the software creator to work well in a particular problem. The bioinformatician should be aware of how well those values apply to the particular problem at hand. Usually, if the problem is similar to the one that motivated the creation of the software the default value will work OK because the original creator of the software usually knows how to optimize his software for that task. When our problem is different from the original one we have to be aware of the changes to adapt the software to our needs.

Gobal and Local Alignments

We could divide the alignment algorithms in two types: global and local. The global algorithms try to create an alignment that covers completely both sequences adding whatever gaps necessary. The local algorithms try to align only the most similar regions. If removing a region from one end of a sequence improves the alignment score they will do it.

Local alignments are usually the best option unless we are sure that the sequences are similar in all its extension. Besides, the local alignment algorithms will create a global alignment with both sequences covered if they are similar enough.

```
Global

  TACGGGGCTAGCTA-TCGTAG

  | | | |      | | |       | | | | | |

  TAGC----TAG----TCGTAG
Local

    TAGCTA-TCGTAG

    | | | | | |  | | | | | |

    TAGCTAGTCGTAG
```

The main practical problem with local alignment algorithms is that they are computationally more demanding that its gobal equivalents. Global alignments are usually only used within the multiple alignment algorithms (alignments with more than two sequences).

Alignment Methods

Dot Plot or Dot matrix

This alignment method creates a graphical representation of the alignment. It creates intuitive representations and it has the advantage that it will show different alternative alignments between two sequences. Other, more standard, alignment methods usually give back only one alignment, the best one, unless instructed otherwise.

Dot plot methods are quite good to study the structure of the sequences involved. They can show repetitions, insertions and deletions clearly.

Once we have identified the regions that match between two sequences we could use another method to create a more conventional text based alignment.

To create a dot plot alignment one sequence is put in the horizontal axis and the other in vertical one. The matches between both sequences are shown as marks in the corresponding position.

```
CATGCT

A   x

T       x   x
```

```
G     x

C  x     x
```

The alignment is shown as a diagonal in the plot.

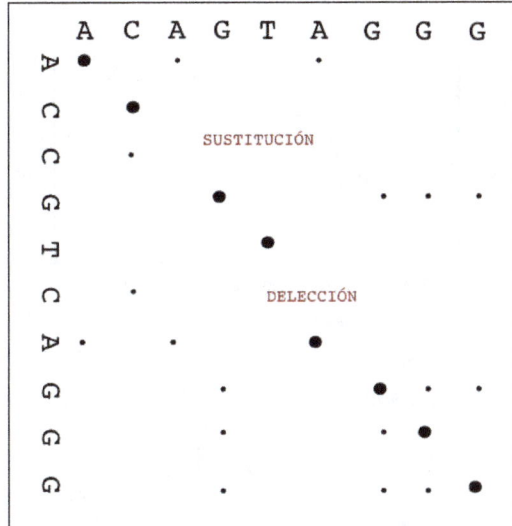

There are different programs to create dot plots. An example is dotmatcher from the emboss suite.

It is easy to detect big insertions, deletions and forward and reverse repeats.

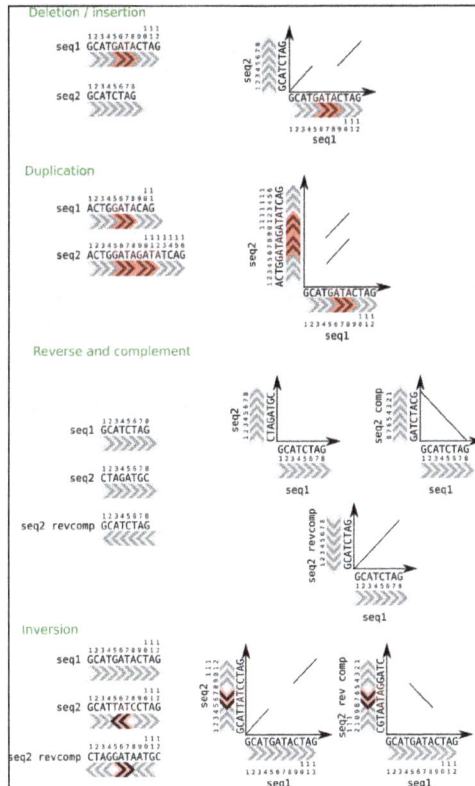

Tandem duplication

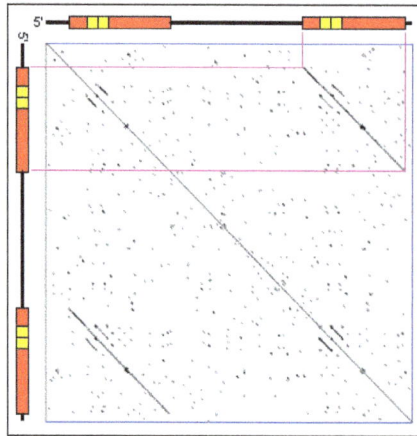

And we can do a genome wide analysis

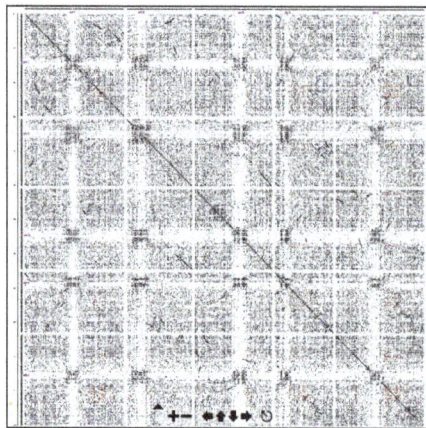

One limitation of the dotplot method is that although we detect similar regions we do not obtain the alignment.

Sensitivity and Specificity

A variation of the dot plot algorithm can compare windows of several residues instead of individual residues. In that case a similarity threshold is set to mark a position as similar. These parameters will influence the sensitivity and specificity of the analysis. If we increase the threshold for a given window size or we decrease the window size for a given threshold we will obtain less spurious signals.

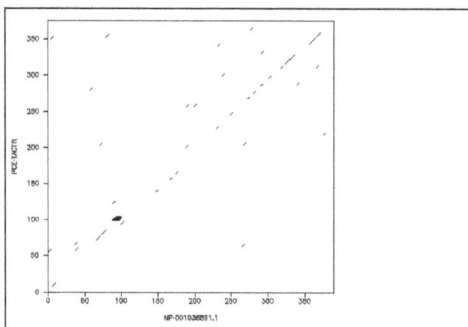

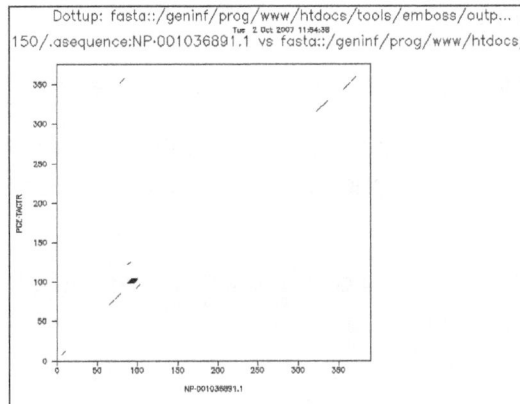

The more stringent the parameters the less noise we will detect, but the more real alignments we will loose. This is related with two very important concepts that can be applied to many bioinformatic and statistical analyses: sensitivity and specificity.

In a dotplot analysis we can draw a dot when two homologous positions match. That would be a true positive. But, if we draw a dot when the two positions are similar but they are not homologous we have a false negative.

With the negatives we have a similar problem. In the dotplot case a negative is a position with no dot. We can have true negatives, positions that does not have a dot and that are not homologous and false negatives, positions that do not have a dot, but that are in fact homologous.

The sentitivity, also called recall, is the proportion of positives that are called positives by the analysis. It is the true positive rate. Ideally we would like to have a 100% percent sensitivity, we would like to mark as positives in the analysis all the true positives.

The specificity is the proportion of true negatives that are called negative by our analysis. It is also know as true negative rate. Ideally we would like to call all true negatives as negatives.

		Truth	
		has the disease	does not has the disease
Test	positive	True Positives (TP)	False Positives (FP)
	negative	False Negatives (FN)	True Negatives (TN)
		senstitivity $\dfrac{TP}{TP + FN}$	**specificity** $\dfrac{TN}{TN + FP}$

In practice most of the time we will miss some positives, so the sensitivity will be lower than 100% and we will detect signals for some negatives. In the dotplot case a false positive would be a region that is marked as similar, but that it is not homologous and a false negative it is a region that is not marked as similar, but that it really is.

Different analysis or different parameter sets will have different sensitivities and specificities. These are very important characteristics of the analyses.

Smith and Waterman Algorithm

There are plenty of algorithms to create text based pairwise sequence alignments.

These algorithms are based on creating a matrix equivalent to the 2D representation created by the dot plot. In those matrices the fragments that will constitute the alignments can be seen as diagonals.

Dotplot diagonals suggest paths in the space of possible alignments. Each of these paths are a possible fragment of the final alignment.

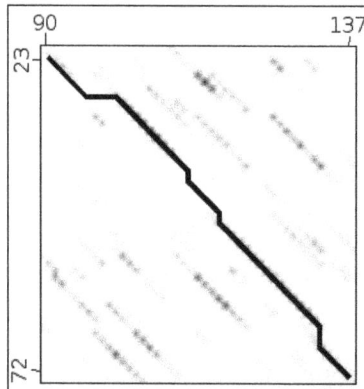

A general approach to detect the path through these diagonals is call dinamic programing. One of the first algorithms implemented following these philosophy was the Needleman-Wunsch algorithm. This algorithm creates global alignments of two sequences given a scoring schema. It is implemented in several programs, one of them is needle from the EMBOSS suite.

```
########################################

# Program:   needle

# Rundate:   Tue Oct 02 10:58:05 2007

# Align_format: srspair

# Report_file: /ebi/extserv/old-work/needle-20071002-10580428983041.output

########################################

#========================================

#

# Aligned_sequences: 2

# 1: SNAK_DROME

# 2: PCE_TACTR

# Matrix: EBLOSUM62

# Gap_penalty: 10.0

# Extend_penalty: 0.5
```

```
#
# Length: 462
# Identity:      126/462 (27.3%)
# Similarity:    184/462 (39.8%)
# Gaps:          114/462 (24.7%)
# Score: 440.5
#
#
#=======================================
SNAK_DROME       1 MIILWSLIVH--LQLTCLHLILQTPNLEALDALEIINYQTTKYTIPEVWK      48
                     ::|:   ..|.|..|::.......|      :.|..::..|
PCE_TACTR        1     MLVNNVFSLLCFPLLMSVVRCSTL------SRQRRQFVFP----      34

SNAK_DROME      49 EQPVATIGEDVDDQDTEDEESYLKFGDDAEVRTSVSEGLHEGAFCRRSFD      98
                                 |||.                        .|...|.
PCE_TACTR       35 ----------------DEEE--------------------LCSNRFT      45

SNAK_DROME      99 GRSGYCILAYQCLHVIREYRVHGTRIDICTHRNNVPVICCPLADKHVLAQ     148
                     ..|.|.....|..::::...:..:..||......|.:||| ...||
PCE_TACTR       46 -EEGTCKNVLDCRILLQKNDYNLLKESICGFEGITPKVCCP-KSSHV---      90

SNAK_DROME     149 RISATKCQEYNAAARRLHLTDTGRTFSGKQCVPSVP-----------LIV     187
                     ||:|:........|         ..||..|::|          .|:
PCE_TACTR       91 -ISSTQAPPETTTTER----------PPKQIPPNLPEVCGIHNTTTTRII     129

SNAK_DROME     188 GGTPTRHGLFPHMAALGWTQGSGSKDQDIKWGCGGALVSELYVLTAAHCA     237
                     ||.....|.:|.|.|.:...||.....|    ||||||:..:|:||:||.
PCE_TACTR      130 GGREAPIGAWPWMTAVYIKQGGIRSVQ-----CGGALVTNRHVITASHCV     174

SNAK_DROME     238 TSGS----KPPDM--VRLGARQLNET--SATQQDIKILIIVLHPKYRSSA     279
                     .:.:    .|.|: |||...|..|  .:...|..:.:.|..:..:.
PCE_TACTR      175 VNSAGTDVMPADVFSVRLGEHNLYSTDDDSNPIDFAVTSVKHHEHFVLAT     224

SNAK_DROME     280 YYHDIALLKLTRRVKFSEQVRPACL----WQLPELQIPTVVAAGWGRTEF     325
                     |.:|||:|.|....|.|::::||.||    .:..:|.:.....|||.|.|
PCE_TACTR      225 YLNDIAILTLNDTVTFTDRIRPICLPYRKLRYDDLAMRKPFITGWGTTAF     274

SNAK_DROME     326 LGAKSNALRQVDLDVVPQMTCKQIYRKERRLPRGIIEGQFCAGYLPGGRD     375
                     .|..|..||:|.|.:.....|:|.|.|:    ..|.....|||:..||:|
```

```
PCE_TACTR         275 NGPSSAVLREVQLPIWEHEACRQAYEKD----LNITNVYMCAGFADGGKD     320

SNAK_DROME        376 TCQGDSGGPIHALLPEYNCVAFVVGITSFGKFCAAPNAPGVYTRLYSYLD     425

                      .|||||||:   :||......:::||.||||.||.|..|||||::..:||

PCE_TACTR         321 ACQGDSGGPM--MLPVKTGEFYLIGIVSFGKKCALPGFPGVYTKVTEFLD     368

SNAK_DROME        426 WIEKIAFKQH         435

                      ||        .:|

PCE_TACTR         369 WI-----AEHMV        375
```

The main limitation of the Needleman–Wunsch algorithm is that it is global, so we should only use it if we know beforehand that both sequences are similar in all its extension.

Smith and Waterman proposed a variation of the algorithm that is capable of generating local alignments. This approach gives better results because it does not force the sequences to be similar in all its extension. If the best, higher scoring alignment, is a global one that aligns completely both sequences the Smith-Waterman algorithm will create it, otherwise it will generate a local alignment.

This algorithm has multiple implementations, one of them is the water program in the EMBOSS suite.

```
#######################################

 # Program:  water

 # Rundate:  Tue Oct 02 11:00:39 2007

 # Align_format: srspair

 # Report_file: /ebi/extserv/old-work/water-20071002-11003873398600.output

#######################################
  #=======================================
  #
  # Aligned_sequences: 2
  # 1: SNAK_DROME
  # 2: PCE_TACTR
  # Matrix: EBLOSUM62
  # Gap_penalty: 10.0
  # Extend_penalty: 0.5
  #
  # Length: 362
  # Identity:     116/362 (32.0%)
```

```
# Similarity:    165/362 (45.6%)

# Gaps:           50/362 (13.8%)

# Score: 452.0

#

#

#=======================================

SNAK_DROME      89 EGAFCRRSFDGRSGYCILAYQCLHVIREYRVHGTRIDICTHRNNVPVICC   138

                   |...|...|. ..|.|.....|..:::::...:.:..||.....|.:||

PCE_TACTR       36 EEELCSNRFT-EEGTCKNVLDCRILLQKNDYNLLKESICGFEGITPKVCC    84

SNAK_DROME     139 PLADKHVLAQRISATKCQEYNAAARRLHLTDTGRTFSGKQCVPSVP----   184

                   | ...||      ||:|:.........|             ..||..|::|

PCE_TACTR       85 P-KSSHV----ISSTQAPPETTTTER---------PPKQIPPNLPEVCG   119

SNAK_DROME     185 -------LIVGGTPTRHGLFPHMAALGWTQGSGSKDQDIKWGCGGALVSE   227

                     .|:||.....|.:|.|.|:...||.....|     ||||||:.

PCE_TACTR      120 IHNTTTTRIIGGREAPIGAWPWMTAVYIKQGGIRSVQ-----CGGALVTN   164

SNAK_DROME     228 LYVLTAAHCATSGS----KPPDM--VRLGARQLNET--SATQQDIKILII   269

                    .:|:||:||..:.:    .|.|: ||||...|..| .:...|..:..:

PCE_TACTR      165 RHVITASHCVVNSAGTDVMPADVFSVRLGEHNLYSTDDDSNPIDFAVTSV   214

SNAK_DROME     270 VLHPKYRSSAYYHDIALLKLTRRVKFSEQVRPACL----WQLPELQIPTV   315

                   ..|..:..:.:.|.:|||:|.|...|.|:::::||.||   .:...:|.:...

PCE_TACTR      215 KHHEHFVLATYLNDIAILTLNDTVTFTDRIRPICLPYRKLRYDDLAMRKP   264

SNAK_DROME     316 VAAGWGRTEFLGAKSNALRQVDLDVVPQMTCKQIYRKERRLPRGIIEGQF   365

                   ...|||.|.|.|..|..|..||:|.|.:.....|:|.|.|:    ..|.....

PCE_TACTR      265 FITGWGTTAFNGPSSAVLREVQLPIWEHEACRQAYEKD----LNITNVYM   310

SNAK_DROME     366 CAGYLPGGRDTCQGDSGGPIHALLPEYNCVAFVVGITSFGKFCAAPNAPG   415

                   |||:..||:|.|||||||: :||......:::||.||||.||.|..||

PCE_TACTR      311 CAGFADGGKDACQGDSGGPM--MLPVKTGEFYLIGIVSFGKKCALPGFPG   358

SNAK_DROME     416 VYTRLYSYLDWI    427

                   |||::..:||||

PCE_TACTR      359 VYTKVTEFLDWI    370
```

Of course, the result will depend not only on the sequences, but on the scoring schema. For instance, if we use a higher penalty for creating gaps we will obtain alignments with fewer gaps.

The nicest property of the Smith-Waterman alignment is that it has been demonstrated that it will generate the optimal alignment, the one with the highest score, given two sequences and a scoring schema. So, if we could, it would be advisable to use always this algorithm. Take into account that several alignments could have the same score, so we can have several alignments with the highest score.

The main problem with the Smith-Waterman algorithm is its slowness. It works very well with small sequences, but it is not practical when the sequences are large. The time it takes to create the scoring matrix for a naive Smith-Waterman implementation depends of m x n (being m and n the length of the sequences).

More than one Alignment

In the dotplot graphical results we saw that some time there could be no just one but several valid alignments. Usually the alignment software that implements the Smith-Waterman algorithm will only print just one alignment by default, the higher scoring one. If there a multiple alignments we will miss them unless we instruct the software otherwise.

Aminoacid Substitution Matrices

When scoring a position of an alignment between two nucleotide sequences we can consider a match if the nucleotide match and a mismatch if the nucleotides are different.

```
ACGT        ACGT

 |

ACGT        ACAT

  ^           ^

 match      mismatch
```

With the aminoacids we can be more subtle because there are aminoacids that are chemically or functionally similar. For instance, we could score higher a substitution of a hydrophobic amino-acid (like Alanine) with another hydrophic aminoacid (like Valine) than with a polar one (like Glutamine).

But if we decide to use such an scoring schema, how should we decided which are the scores to use for each possible aminoacid substitution? A possible way of creating such an aminoacid subtitution scoring matrix would be to count how many times each pair the possible aminoacid substitutions are found at homologous positions in alignments of homologous proteins. The pairs that tend to appear at the same positions could be considered functionally equivalent and scored higher than the ones that in few instances are found at the same positions. Following this method we could create a substitution matrix for all possible aminoacid substitutions.

A series of matrices built in such a way are the PAM matrices. There are several PAM matrices and not just one because they are built from comparison of sequences that are closer or further appart in evolutionary time. They are named by the number of aminoacid mutations for every 100 ami-noacids that differentiate the sequences compared to create the matrices. There are, for instance

PAM100, PAM160 and PAM200 matrices. Take into account that a substitution can underlie several mutations, hence there could be more than 100 mutations in 100 aminoacids. A higher number is related with a longer evolutionary time.

Another commonly used set of matrices are the BLOSUM matrices. They are based on the same idea, but instead of being built with complete alignments they are built by using highly conserved blocks. This allows the BLOSUM matrices to cover longer evolutionary times than the PAM matrices because aligning very distant sequences in all their extension can be challenging, but aligning the most conserved blocks of those sequences could be easier. In the BLOSUM matrices the number reflects the minimum percentage of identity allowed between the sequences so BLOSUM 70 used sequences that were more similar than BLOSUM 50.

We will see that a very related concept, the nucleotide evolution model, is used in the phylogenetic analyses, but they are seldom used by the alignment software.

Statistical Significance of the Alignments

One general problem with the bioinformatic algorithms is that they usually generate a result, but some times that result could be meaningless. For instance, we could generate two random sequences and align them. The Smith-Waterman algorithm would generate an alignment, but despite being its optimal alignment it would be meaningless.

In the DNA alignments is quite easy to decide with alignments are meaningful because the spurious ones tend to be very short. In the protein alignments it is usually not that clear because we are allowing to aminoacid to be similar and not just identical and that tends to produce longer alignments even when they are meaningless.

The solution to this problem is to calculate some statistic that reflect the significance of the solution obtained. As we will see there is software that calculate these kind of statistical significance measures, like BLAST.

Multiple Sequence Alignment

A Multiple Sequence Alignment is an alignment of more than two sequences. We could align several DNA or protein sequences.

Some of the most usual uses of the multiple alignments are:

- Phylogenetic analysis.

- Conserved domains.

- Protein structure comparison and prediction.

- Conserved regions in promoteres.

The multiple sequence alignment asumes that the sequences are homologous, they descend from a common ancestor. The algorithms will try to align homologous positions or regions with the same structure or function.

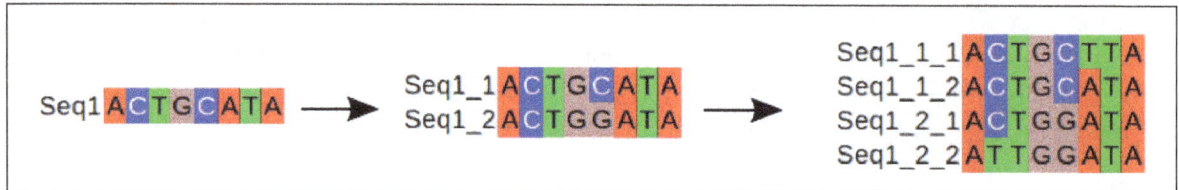

Multiple Alignment Algorithm

Multiple alignments are computationally much more difficult than pair-wise alignments. It would be ideal to use an analog of the Smith & Waterman algorithm capable of looking for optimal alignments in the diagonals of a multidimensional matrix given a scoring schema. This algorithm would had to create a multidimensional matrix with one dimension for each sequence. The memory and time required for solving the problem would increase geometrically with the lenght of every sequence. Given the number of sequences usually involved no algorithm is capable of doing that. Every algorithm available reverts to a heuristic capable of solving the problem in a much faster time. The drawback is that the result might not be optimal.

Usually the multiple sequence algorithms assume that the sequences are similar in all its length and they behave like global alignment algorithms. They also assume that thre are not many long insertions and delections. Thus the algorithms will work for some sequences, but not for others.

These algorithms can deal with sequences that are quite different, but, as in the pair-wise case, when the sequences are very different they might have problems creating good algorithm. A good algorithm should align the homologous positions or the positions with the same structure or function.

It we are trying to align two homologous proteins from two species that are phylogenetically very distant we might align quite easily the more conserved regions, like the conserved domains, but we will have problems aligning the more different regions. This was also the case in the pair-wise case, but remember that the multiple alignment algorithms are not guaranteed to give back the best possible alignment.

These algorithms are not design to align sequences that do not cover the whole region, like the reads from a sequencing project. There are other algorithms to assemble sequencing projects.

Progressive Contruction Algorithms

In Multiple Sequence Alignment it is quite common that the algorithms use a progressive alignment strategy. These methods are fast and allow to align thousands of sequences.

Before starting the alignemnt, as in the pair-wise case, we have to decide which is the scoring schema that we are going to use for the matches, gaps and gap extensions. The aim of the alignment would be to get the multiple sequence alignment with the highest score possible. In the multiple alignment case we do not have any practical algorithm that guarantees that it going to get the optimal solution, but we hope that the solution will be close enough if the sequences comply with the restrictions assumed by the algorithm.

The idea behind the progressive construction algorithm is to build the pair-wise alignments of the more closely related sequences, that should be easier to build, and to align progressively these alignments once we have them. To do it we need first to determine which are the closest sequence pairs. One rough and fast way of determining which are the closest sequence pairs is to align all the possible pairs and look at the scores of those alignments. The pair-wise alignments with the highest scores should be the ones between the more similar sequences. So the first step in the algorithm is to create all the pair-wise alignments and to create a matrix with the scores between the pairs. These matrix will include the similarity relations between all sequences.

Once we have this matrix we can determine the hierarchical relation between the sequences, which are the closest pairs and how those pairs are related and so on, by creating a hierarchical clustering, a tree. We can create these threes by using different fast algorithms like UPGMA or Neighbor joining. These trees are usually known as guide trees.

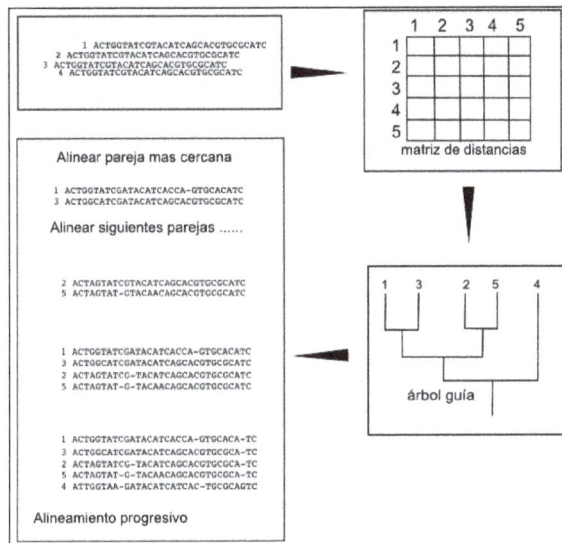

An example:

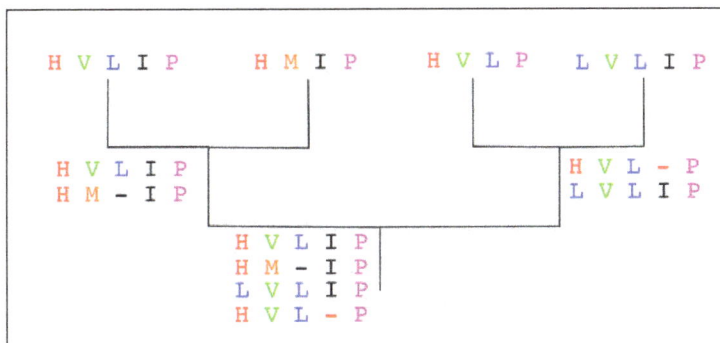

Another example:

```
Secuences:

    IMPRESIONANTE

    INCUESTIONABLE

    IMPRESO

    Scores:

        IMPRESIONANTE X IMPRESO 7/13

        IMPRESIONANTE X INCUESTIONABLE 10/14

        INCUESTIONABLE X IMPRESO 4/14

    Scoring pair-wise matrix:
```

	IMPRESIONANTE	INCUESTIONABLE	IMPRESO
IMPRESIONANTE	1	10/14	7/13
INCUESTIONABLE	10/14	1	4/14
IMPRESO	7/13	4/14	1

```
    Guide Tree:

            |--- IMPRESIONANTE

      |---|--- INCUESTIONABLE

      |

      |----- IMPRESO

The first alignment would be: IMPRESIONANTE x INCUESTIONABLE

    IMPRES-IONABLE

    INCUESTIANABLE

Now we align IMPRESO to the previous alignment.

    IMPRES-IONANTE

    INCUESTIONABLE

    IMPRES--O-----
```

We have no guarantee that the final is the one with the highest score.

The main problem of these progressive alignment algorithms is that the errors introduced at any point in the process are not revised in the following phases to speed up the process. For instance, if we introduce one gap in the first pair-wise alignment this gap will be propagated to all the following alingments. If the gap was correct that is fine, but if it was not optimal it won't be fixed. These methods are specially prone to fail when the sequences are very different or phylogenetically distant.

```
Sequences to align already in the order given by a guide tree:

Seq A  GARFIELD THE LAST FAT CAT

Seq B  GARFIELD THE FAST CAT

Seq C  GARFIELD THE VERY FAST CAT

Seq D  THE FAT CAT

Step 1

Seq A  GARFIELD THE LAST FAT CAT

Seq B  GARFIELD THE FAST CAT

Step 2

Seq A  GARFIELD THE LAST FA-T CAT

Seq B  GARFIELD THE FAST CA-T

Seq C  GARFIELD THE VERY FAST CAT

Step 3

Seq A  GARFIELD THE LAST FA-T CAT

Seq B  GARFIELD THE FAST CA-T

Seq C  GARFIELD THE VERY FAST CAT

Seq D  -------- THE ---- FA-T CAT
```

Historically the most used of the progressive multiple alignment algorithms was CLUSTALW. Nowadays CLUSTALW is not one of the recommended algorithms anymore because there are other algorithms that create better alignments like Clustal Omega or MAFFT. MAFFT was one of the best contenders in a multiple alignment software comparison.

T-Coffee is another progressive algorithm. T-Coffee tries to solve the errors introduced by the progressive methods by taking into account the pair-wise alignments. First it creates a library of all the possible pair-wise alignments plus a multiple alignment using an algorithm similar to the CLUSTALW one. To this library we can add more alignments based on extra information like the protein structure or the protein domain composition. Then it creates a progressive alignment, but it takes into accounts all the alignments in the library that relate to the sequences aligned at that step to avoid errors. The T-Coffe algorithm follows the steps:

- Create the pair-wise alignments.

- Calculate the similirity matrix.

- Create the guide tree.

Build the multiple progressive alignment following the tree, but taking into account the information from the pair-wise alignments

T-Coffee is usually better than CLUSTALW and performs well even with very different sequences, specially if we feed it more information, like: domains, structures or secondary structure. T-Coffee is slower than CLUSTALW and that is one of its main limitations, it can not work with more than few hundred sequences.

Iterative Algorithms

These methods are similar to the progressive ones, but in each step the previous alignments are reevaluated. Some of the most popular iterative methods are: Muscle and MAFFT are two popular examples of these algorithms.

Hidden Markov Models

The most advanced algorithms to date are based on Hidden Markov Models and they have improvements in the guide tree construction, like the sequence embedding, that reduce the computation time.

Clustal Omega is one of these algorithms and can create alignments as accurate of the T-Coffee, but with many thousands of sequences.

Alignment Evaluation

Once we have created our Multiple Sequence Alignment we should check that the result is OK. We could open the multiple alignment in a viewer to assess the quality of the different regions of the aligment or we could automate this assesment. Usually not all the regions have an alignment of the same quality. The more conserved regions will be more easily aligned than the more variable ones.

It is quite usual to remove the regions that are not well aligned before doing any further analysis, like a phylogenetic reconstruction. We can remove those regions manually or we can use an especialized algorithm like trimAl.

Software for Multiple Alignments

There are different software packages that implement the described algorithms. These softwares include CLI and GUI programs as well as web services.

One package usually employed is MEGA. MEGA is a multiplatform software focused on phylogenetic analyses.

Jalview and STRAP a multiple alignment editor and viewer. Another old software, that has been abandoned by its developer is BioEdit.

In the EBI web server have some services to run several algorithms like: Clustal Omega, Kalign, MAFFT, and Muscle.

Bloom Filters for Bioinformatics

The Bloom filter was originally developed by Burton H. Bloom back in the seventies and for long time it was there without any major application. To making Bloom filter popular again. Only after the Google used Bloom filters for their BigTable database system, the idea started grabbing the attention of larger and diverse audience.

Bloom filter is an extremely space-efficient probabilistic data structure which enables the quick and easy membership tests. On positive side, in order to test whether or not an element is a member of a set, Bloom filters need less memory than any other data structure such as hash tables, simple arrays or linked lists. On downside, the risk of false positives is higher.

Any task that require checking two sets of data against each other can be performed in an efficient way using Bloom filter. For instance, one can use Bloom filter to do spell-checking against a dictionary of correct words.

Recently Bioinformatics community started using Bloom filters for large scale gene sequence analysis. Comparing sequences to test similarity is a common task in bioinformatics.

Stranneheim et. al describe a novel Bloom filters based algorithm, FACS (Fast and Accurate Classification of Sequences) for accurate and rapid classification of DNA sequences as belonging or not belonging to a reference DNA sequence. In this case reference DNA sequence can be as large as the whole genome. This kind of rapid classification method can be a boon for metagenomic studies where one need to classify sequences as 'novel', or belonging to a well known genome.

A comparative study using metagenomic data sets suggest that FACS method is at least 21 times faster compared to algorithms such as BLAT and SSAHA2 with nearly same accuracy. The FACS algorithm is implemented as PERL module and it can be downloaded from CPAN.

Similarly Malde and O'Sullivan have developed some interesting bloom filter based sequence analysis applications in Haskell. In their analysis they matched randomly selected ESTs against the E. coli genome which is relatively small compared to human genome.

In terms of memory consumption their Bloom filter application was using a mere 20MB, of which the Bloom filter itself needed only 2MB compared to the set based implementations those allocated 160-190MB of memory for a small test case.

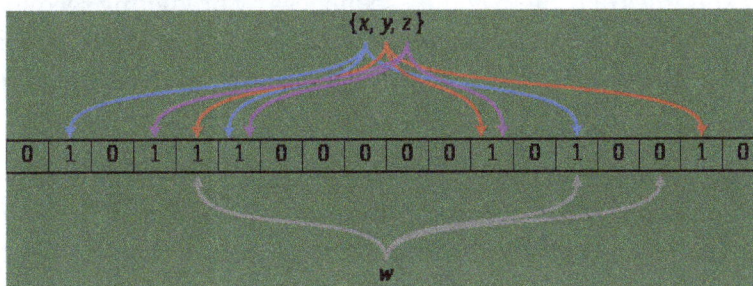

Bloom Filter

Dot Plots

Dot plots are a data analysis tool to scrutinise symbol sequences. Arrangement recurrences, similarities and inherent structures are visualised in two-dimensional plots. Repetitions of individual symbols, as well as motifs being similar appear as characteristic lines, rectangles, textures or combinations thereof. Through visualisation such structures can intuitively be distinguished from background noise and conceived by human eye. For two arbitrary sequences dot plot algorithms completely enumerate substrings of defined length. Each substring of the first sequence is aligned with each of the second. For each alignment a score for pairwise matching symbols among the two substrings is computed and visualised by colour mapping. The substring positions in the sequence determine the position in the dot plot and result in a concise representation of similarities among sequences.

Background

In the field of genetics, the tools most commonly used today are alignment-based methods. However, alignment-based methods generally interpret incongruent character states in relation to surrounding identities, as the result of punctual deletion, insertion and alteration of character state. Consequently, they tend to perform poorly when the actual evolutionary process violates these implicit assumptions, e.g. when reversal or relocation of parts of a sequence took place. In addition, by way of their quest to present a single solution of positional homology, alignment algorithms are not designed to identify simple sequence repeats. To address this task dot plots are a powerful approach. Beside perfect motif repetition dot plots also reveal particular information on imperfect repetitions.

The dot plot technique dates back to the late 1960's and early 1970's and was developed in the field of genetics. Today, the most powerful application for dot plots is still the domain where they initially emerged from. Since the early approaches more sophisticated software tools have been designed and continuing technological progress allows for increasingly faster and more complex analyses. Several state of the art programs have implemented dot plot modules for homology comparisons and pattern recurrence detection. Dot plot applications are particularly useful in the identification of interspersed repeats such as transposons and tandem-repeat motifs such as microsatellites. Furthermore, loss or gain of whole motifs can easily be spotted in different types of domains, a trait useful in characterising the evolution of certain protein families. Dot plots are also employed in the investigation of properties of protein coding sequences by predicting secondary structures, like stem-loop formation or structural RNA domains. In the field of DNA based watermarks adopted algorithms can support identification of patent infringements of genetically modified organisms during routine screening.

Although the main application of dot plot lays in genetics the method itself is context independent. Therefore it can be applied to any sequences where symbols represent items or tokenised objects. Thus, the method allows to investigate patterns in texts or chronological events and has a broad application field when one requires scrutinising sequences for such patterns. Dot plots have also been applied in software development, to identify redundant code and to ease maintenance of design patterns among releases. Historians and literary scientists can use dot plots to follow transcriptions and semantics over time. Comparing sequences of different origin they allow visual identification of source insertions, deletions or homologies, culminating in plagiarism detection. In highly volatile text systems, like websites, they have proven to support tracking of evolutionary modification processes. In translations they allow a more effective organisation of workflows and help institutions to reduce expenses.

Algorithmic Implementation

Initial Example

As an initial example for dot plots one can imagine the same sequence written onto two strips of chequered paper. Every symbol of the sequence is written consecutively into one chequer, with its index number next to it. By overlaying a frame containing a window that allows viewing exactly one symbol of each strip at a time symbols are compared in pairs. Whenever symbols in the observing windows match, a bright dot is placed in a grid at the respective indices. The resulting rectangular graphical representation is a dot plot. It thus represents all possible comparisons of characters in either sequences and is colour-coded with two colours indicating a match or mismatch between any two characters. The resulting rectangular graphical representation is a dot-plot.

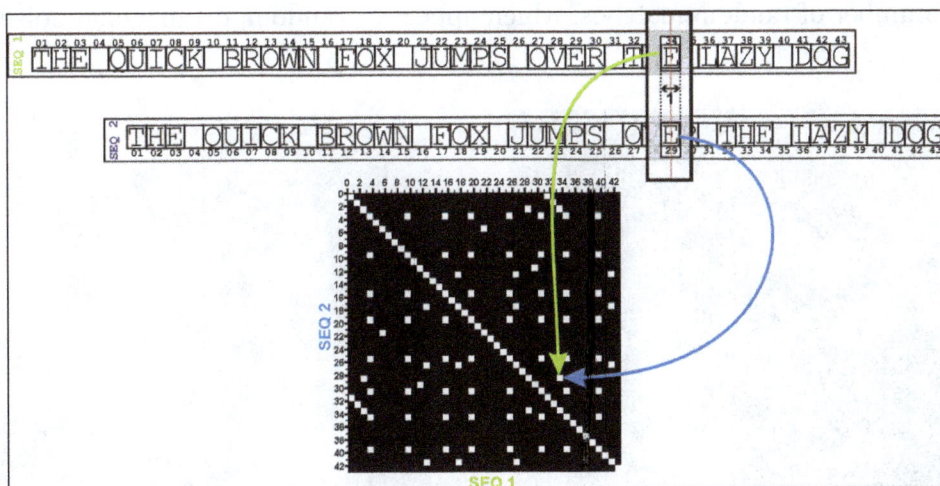

A generalised description of how to create a simple dot plot based on two copies of a sequence with an underlying alphabet of 27 different symbols (26 symbols plus space) written onto two paper strips. By moving the two strips SEQ1 and SEQ2 under the frame with a window of the width of one symbol, it is possible to iterate every possible pairing. Every time the symbols in the window of the frame match, a white dot is placed at the respective indices. Conversely, a black dot is placed on the grid when symbols mismatch.

The dot plot result of comparing all symbols with each other can be represented by a matrix hereafter named M. The two sequences are furthermore referred to as T1 and T2. In this initial example is T1=T2. In the following we respect the formal convention that the first index refers to the row, the second the column; M(row, column). As the first sequence is normally presented on the horizontal axis processing of T1 is assigned to column and T2 to the row indices.

Detection of Signal and Noise in Dot Plots

Here, the sequence was compared against itself and results in a self-similarity dot-plot. The emerging dot plot shows a pronounced diagonal with a symmetric distribution of several points on both sides of it. The diagonal marks the comparison of each symbol with itself and thus appears as a distinct and continuous line. The recurrent appearance of the motif 'THE' followed by a space is a second pattern creating a four dots long line. As this repetition has the same reading direction within the sequence, the line is parallel to the diagonal. The horizontal and vertical offset along the axes is equal to the respective distance in the sequence. In addition to recurrences of individual symbols whole motifs can be found as reverted copies within larger sequences and result in at least partially palindromic structures. The term palindrome defines motifs of individual symbols that read the same from both directions.

Single bright dots in the plot indicate positions where other symbols match. As these points are scattered on both sides of the diagonal and show no regular organisation, they can be assumed to represent random matches. Random matches are a function of sequence length and alphabet size. The stochastic probability of random pattern repetitions decreases with increasing alphabet sizes, while the total number increases with sequence length. In this example a relatively short sequence with a large alphabet of 27 symbols (characters 'A' to 'Z' and the space character) was used. This reduces the number of random matches, which appear as random, off-diagonal noise in the dot plot.

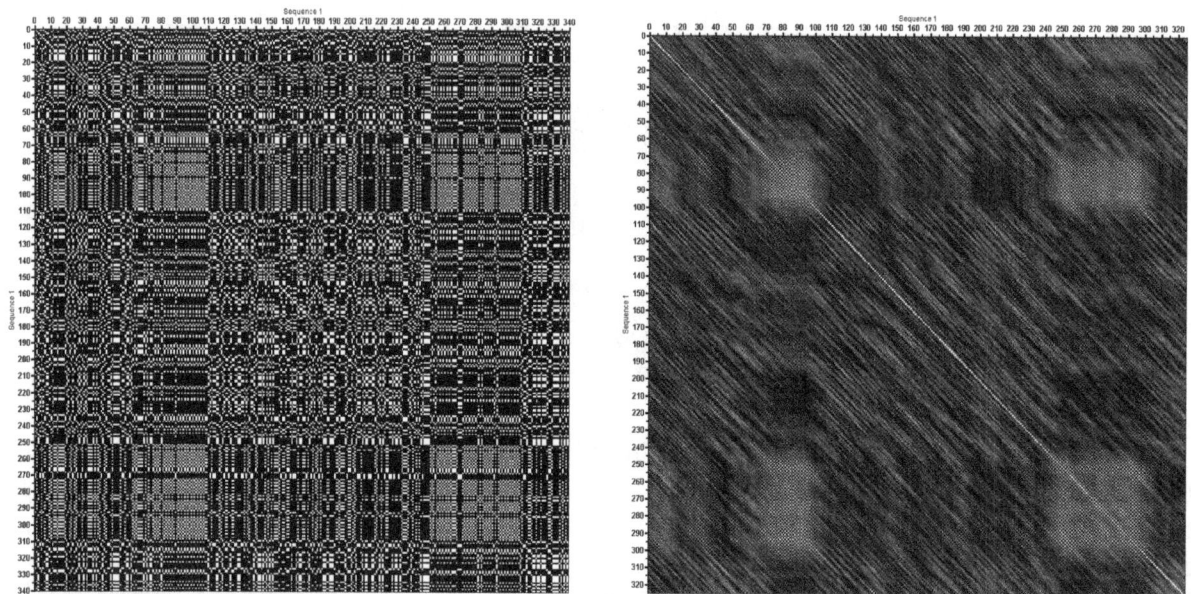

Two identity dot plots of the DNA sequence EU127468.1 compared against itself. Upper panel: Char-by-char comparison (window size w = 1) and two colours. Lower panel: Window size of w = 16 characters and a linear colour mapping. Note different scaling due to fewer alignments

Investigating sequences of nucleic acids an alphabet size of four symbols and sequence lengths of several hundreds to thousands characters are common. This entails more background noise as compared to a dot plot of amino acid sequences. The probability of a random match between any two nucleic acids is 0.25 whereas it is 0.05 for amino acids with an alphabet size of 20. Using a window size of one symbol for the analysis of the arbitrary gene sequence EU127468.1 the plot has a high level of background noise and is of minor use, when interested in large scale patterns or structures. This effect is even more pronounced, when applying dot plots to sequences of binary information. Consequently, filters are required to emphasise regions with a number of consecutively matching symbols. Therefore window size is enlarged to compare pairs of more than one consecutive symbols for each dot in the plot.

Table: Genetic microsatellite sequence EU127468 of the marine, benthic isopod Serolis paradoxa Spa04. Two regions with an imperfect (AC)x microsatellite are printed in bold. Sequence is broken into pieces of 40 nucleic acids.

```
GCTGAGCTTACGAACAAAACTGCAGCAGTGATGTAAATAT

ACGTACGGTATTTTAGGCTTGTACACCCCTCTATTACACA

TACACACACGCACACACACACACACACACCTGAGGTTACT

GAAGTAAGGTTGGAGACGGTACTTGTCTATCTCCCAGCCG

AAGTGGTCTTCCGCTGAGCAGAGTTCCTTTGCCACCCTGA

ATCATGGCTGTTGGTTCAGTATGAAGGTGTTGTACCCAGT

AGCGCTGTTTTACAGACACACACACAAACGGGCACACACA

CATATACACACAGATACACACACACACTAGGTAACTTA

TTCTTTGAACTTTTCTATCT
```

Identity dot plots compare substrings of defined length starting from every position on the respective strands. As alignments within the windows are only performed in reading direction the detection of reverted structures and palindromes is reduced with window sizes larger than one. Beside literal palindromes palindromic structures also play an important role in genetic sequences, where definition is slightly different. The DNA double helix is formed by two complementary paired strands. Biological reading direction follows the chemical enumeration of the phosphate linked ribose backbone from the 5' to 3' end and is oppositely oriented on the two strands. The base pairing between strands is determined by the two complementary base pairs adenine and thymine (A-T), as well as cytosine and guanine (G-C). Such sequences, or parts of it are palindromic when the first strand reads the same as its reverse complement in the respective reading direction. To overcome limitations of the identity dot plot, a parallel analysis needs to be included, to detect structures like the motif 5'-TATAGCTATA-3', which is palindromic to its reverse complement. Thus, the detection of reverted motifs on the complement strand in particular for larger window sizes is of major interest. Therefore sequence are also compared with reverted copies or in case of a genetic sequence with its reverse complement of the opposite strand.

Colour Mapping

The elements of the result matrix M are ordinal numbers representing the number of pairwise matches for the respective substrings. These ordinals can be directly used as indices pointing at entries in a colours palette array. The size of the colour map is the maximum number of possible matches plus one colour for the background; equivalent to #colour = 1+w. By using a grey-scale palette areas appear gradually highlighted by their number of consecutively matching symbols. Additionally, one colour can be assigned to more than one element of the colour map and thus to more than one ordinal. Black is often used for a range of low numbers, to fade out regions of low similarity or complexity. Assigning bright colours to more ranks than the highest these structures are emphasised.

Features of Window Size

The chosen window size is of eminental importance in the visualisation process as it determines the criteria by which imperfect matches are distinguished as geometric figures from the background. Larger values allow finer colour graduations and focus on larger structures of consecutively matching symbols. In palindromes of even length each symbol appears at least twice, while odd lengths may have a unique central character. Thus, window sizes of one symbol impede identification of such patterns. For dot plots including reverted analyses appropriate window sizes are generally larger than three.

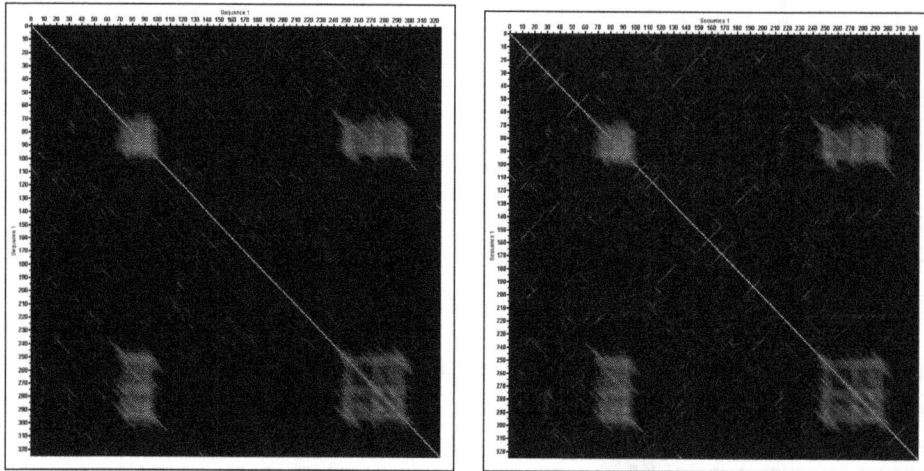

Figure: Two dot plots of the DNA sequence EU127468.1 compared against itself.

Window size for both plots is $w=16$ and colour mapping assigns black to cells where counts are 7 or less. Upper panel: Identity dot plot. Lower panel: Dot plot including the reverse complement. Comparing upper and lower panel it can be seen that also similarities with the reverted strand exist.

Features of Algorithms

Comparing two different sequences the dot plot matrix needs to be entirely calculated. Investigating the algorithms it is found that runtime is a function of $l*k*w$ for source code 1 and $l*k*2w$ for source code 2. It is easily seen that runtime increases with increasing w, as often required for already larger sequences. Compared to fast pattern matching algorithms, like Knuth-Morris-Pratt or Boyer Moore, dot plots have the disadvantage that symbols within the windows need to be evaluated completely, but have the major advantage that they allow distinguishing similarities from imperfections. Thus, optimisation of match evaluation in the sliding window is a limiting factor. Sonnhammer and Durbin introduced a method with vectors, initialised for pairwise matches whenever the sliding window moves over a matching position in the sequences. Each vector contributes to the result as long as its position lays within the window. The outermost vector is discarded, when the window shifts one position towards the next position to compare. Huang & Zhang introduced a method creating look-up tables and initial tokenising of possible symbol combinations to increase calculation speed for DNA/RNA dot plot applications. The initial table setup requires time paying off with larger sequences. Small alphabets of the four nucleotides agree with this method. However, tables grow exponentially when forming words of certain lengths including e.g. the complete set of DNA IUPAC symbols.

To reduce memory consumption for larger sequences compressions are used to focus on large scale structures. Therefore the plot is split into rectangles of equal size. Within these rectangles scores are evaluated completely. In the compressed plot each rectangle is represented by a single dot giving a proxy value for the rectangle. The disadvantage is that calculations need to be performed twice when zooming into a plot.

To reduce stack load results of substring comparison can be directly passed to a visualisation routine and not returned as complete l´k matrix. When a sequence is compared with itself less than half of the result matrix needs to be calculated. In this case it can be made use of the fact that $M(i, j) = M(j, i)$ and $M(i, i) = w$.

Comparing a sequence simultaneously with a couple of others it is possible to overlay results. To identify common similarities further methods are available to analyse dot plots for motif and alignment recognition and to classify sequences by multivariate analyses of dot plots.

For some applications it may also be interesting to create dot plots that display lowest instead of highest counts to visualise matches being different from zero. Additionally one might be interested in different ways of counting or evaluating sequence substrings, to weigh matches differently or to return a normalised substring similarity ($q\% = 1 / w *$ #matches) for subsequent numerical processing.

Interpreting Dot Plots

The upper left corner of a dot plot normally shows the comparison of the first substring set from the two sequences. Down- and rightward indices of alignments increase. Especially in older publications origins of sequences is sometimes found in the lower left corner, increasing up- and rightward. The operating direction of a software package can be easily identified by plotting a sequence against itself. In this case dots on the main diagonal represent substrings from the positions $i=j$. This is equal to compare a substring with itself and results in a pronounced main diagonal where the number of matches is equal to the windows size w (figure a). The plot is then square and symmetric across the main diagonal. Patterns off the main diagonal indicate structural similarities among different parts of the sequences. Lines off the main diagonal indicate that the different regions on the sequences are have a high similarity, maybe due to repetition (figure b). The location of these repetitions is equal to the axis intercepts in the plot (D_1, and D_2 in figure b). When D_1 and D_2 intersect on the same axis, regions in the respective strand are present that bear self-similarities. Including the reverted or complementary strand in the analysis, sequence similarities of reverted motifs are indicated by lines orthogonal to the main diagonal (c, d). In genetics many transposable elements are characterised by short duplicated regions or terminal inverted sequence repeats that can be identified by their typical structure in dot plots (figure d). Tandem repeats, such as microsatellites (figure e) and minisatellites (microsatellites = repeat unit 1-6 base pairs, minisatellites = repeat unit of 10-50 base pairs) are characterised by parallel lines forming a rectangular shape. If two different sequences are compared against each other (figure g, h) gaps in the main diagonal indicate deviations. If the main diagonal shows a break (figure h) this point indicates the location of a fragment in one strand that was deleted or inserted in the other one (so called 'indel'). This is also a common feature when plotting cDNA sequences of a gene against the unspliced genomic sequence.

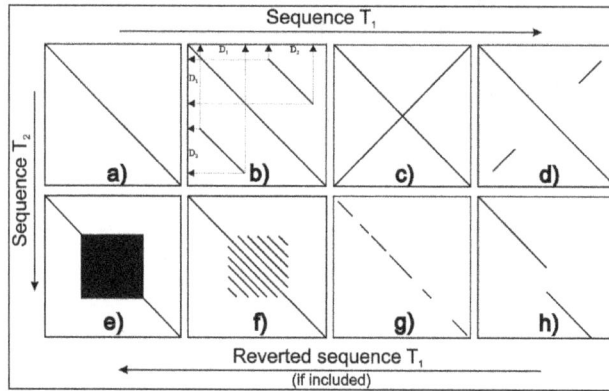

Figure: Schematic overview of characteristic patterns appearing in dot plots. a-f) are self similiarity dot plots (T1=T2). g-h) are dot plots comparing two different sequences of simlar length.

a) A continuous main diagonal shows perfect similarity for symbols with the same indices.

b) Parallels to the main diagonal indicate repeated regions in the same reading direction on different parts of the sequences. In this case a region D is found twice in the sequence (D_1, D_2, so called 'duplications').

c) Lines perpendicular to the main diagonal indicate palindromic areas. In this case the sequence is completely palindromic in the displayed area. As an example the latin sentence 'sator arepo tenet opera rotas' might be consulted.

d) Partially palindromic sequence (For DNA sequences this refers to a perfect match of the normal strand with its reverse complement, which is frequently found for many transposable elements.

e) Bold blocks on the main diagonal indicate repetition of the same symbol in both sequences, e.g. (G)50, so called microsatellite repeats.

f) Parallel lines indicate tandem repeats of a larger motif in both sequences, e.g. (AGCTCTGAC)20, so called minisatellite patterns. The distance between the diagonals equals the distance of the motif.

g) When the diagonal is a discontinuous line this indicates that the sequences T1 and T2 share a common source. In literal analyses we may have to deal with plagiarism or in DNA analyses sequences may be homologous because of a common ancestor. The number of interruptions increases with modifications on the text or the time of independent evolution and mutation rate.

h) Partial deletion in sequence 1 or insertion in sequence 2, so called 'indel'. In protein coding sequences this can be often observed for many different types of domains, which got lost or substituted during evolution. Also comparing mRNA (cDNA) sequences without introns (T_1) against the unspliced DNA sequence (T_2) generally yields this picture.

Analysing the identity dot plot ($T_1=T_2$) of the sequence EU127468.1 the continuous main diagonal is found. Additionally, four bright rectangular patterns can be seen. Two of them are square and located on the main diagonal approximately between positions 65-95 and 245-290. They represent regions of high self similarity due to local microsatellite repetition of short (AC)x tandems. The rectangular patterns off the diagonal indicate similarity between the two repeat regions on the diagonal to which they are located perpendicular. This results from high similarity that is not only found around position $T_1[i]/T_2[i]$ and $T_1[j]/T_2[j]$ but also between $T_1[i]/T_2[j]$ and $T_1[j]/T_2[i]$ and makes up the four symmetric rectangles recognized. the lengths of the sides of the rectangles are equal to the outlines of the opposite squares on the diagonal and interconnected with these by broad dark streaks. They indicate the reduced probability of random matches between the tandem repeats and non-repeat regions.

Application of Dot Plot in Genetic Research

A concise application for dot plots is genetic probe design and comparsion of evolutionary altered sequences. Here, the advent of the polymerase chain reaction paved the way for a variety of applications that rely on identifying and amplifying specific genomic regions. Selected probes often fail to amplify the specified regions or amplify multiple genomic regions. While several probe-design programs aid the process of design and optimisation according to thermodynamic rules, amplification can fail as probes might hybridise to regions in the complementary strand unnoticed by the primer design program, as the motif is completely different. Dot plots including reverse complement analysis allow identification of such sites. For microsatellite analyses in population genetic or mapping studies the aim is to specifically amplify highly variable genomic regions using conserved flanking regions. Flanking regions are often much less conserved and an integral part of higher-order repeats. Dot plots allow the identification of these super-order structures and avoid locating probes within them.

If a priming site is duplicated or even duplicated further obscured due to inversions, successful and specific hybridization or amplification of single sites can be counteracted. High numbers of sequences from the enrichment procedure can be screened. While appropriate tools for the identification of tandem repeats exist the detection of higher-order or highly degenerated repeats, which should be avoided as candidate markers, still pose a complicated issue. Additionally, appropriate primer sites should not be located within duplicated regions, counteracting amplification success or leading to the amplification of paralogous gene copies. Dot plots greatly facilitate testing for the presence of such sites and proved to be helpful tools for the de novo development of appropriate microsatellite markers.

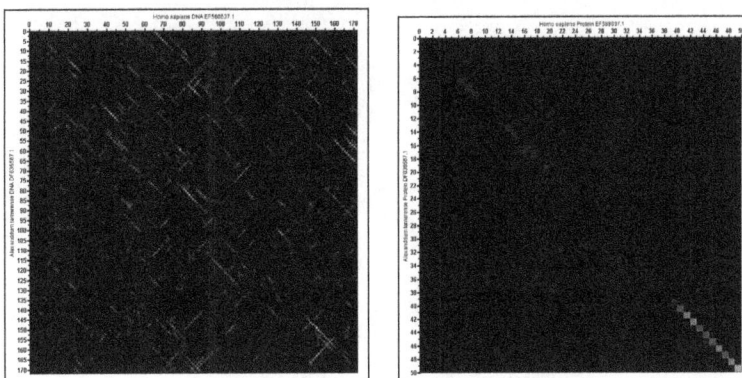

Figure: Dot plot of a partial sequence of the cytochrome c oxidase, subunit 1 of *Homo sapiens*
(T_1, GenBank accession number EF568637) against that of the unicellular dinoflagellate
Alexandrium tamarense (T_2, GenBank accession number EF036567).

Window size for both plots is $w=12$ and colour mapping assigns black to cells where counts for pairwise substring matches are 5 or less. While only few conspicuous similarities are recognized on DNA level (upper panel), shared similarity is much more conserved on protein level (lower panel).

Comparative genomic and phylogenetic studies compare DNA regions with the same ancestry and analyse the amount of accumulated differences. With growing phylogenetic distance differences increase, so that inherited similarities are difficult to detect using standard alignment methods. This holds true in particular if rearrangements of sequence domains have occurred. Here, dot plots are a powerful tool to visualise similarities and deviations. Cytochrome c oxidase is a small protein in mitochondrial membranes. Comparing a partial DNA sequence from Homo sapiens with that of the distantly related dinoflagellate Alexandrium tamarense few similarities are found. The resulting dot plot, including reverse complement analysis, shows a disrupted main diagonal as a result of several substitutions. Additionally several motifs exist, which are palindromic in the opposite sequence. However, when the translated amino acid sequences is compared a much higher conservation level is found due to the degenerated genetic code at DNA level.

Examples and Interpretations of Dot Plots

Contrary to simple sequence alignments dot plots can be a very useful tool for spotting various evolutionary events which may have happened to the sequences of interest.

Below is shown some examples of dot plots where sequence insertions, low complexity regions, inverted repeats etc. can be identified visually.

Similar Sequences

The most simple example of a dot plot is obtained by plotting two homologous sequences of interest. If very similar or identical sequences are plotted against each other a diagonal line will occur.

The dot plot in two related sequences of the Influenza A virus nucleoproteins infecting ducks and chickens. Accession numbers from the two sequences are: DQ232610 and DQ023146.

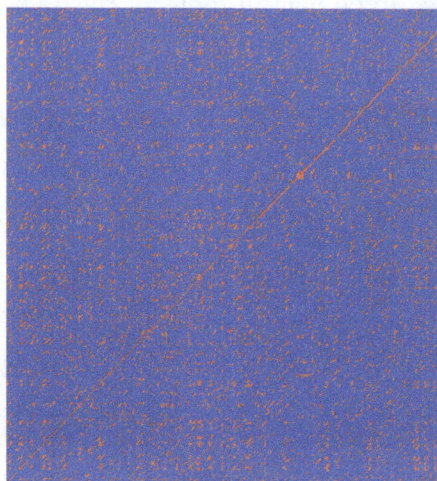

Figure: Dot plot of DQ232610 vs. DQ023146 (Influenza A virus nucleoproteins) showing and overall similarity

Repeated Regions

Sequence repeats can also be identified using dot plots. A repeat region will typically show up as lines parallel to the diagonal line.

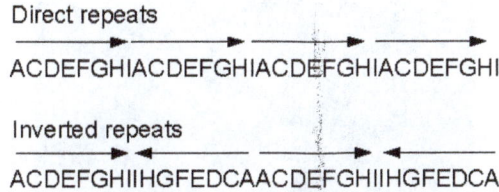

Direct repeats

ACDEFGHIACDEFGHIACDEFGHIACDEFGHI

Inverted repeats

ACDEFGHIIHGFEDCAACDEFGHIIHGFEDCA

Figure: Direct and inverted repeats shown on an amino acid sequence
generated for demonstration purposes.

If the dot plot shows more than one diagonal in the same region of a sequence, the regions depending to the other sequence are repeated. In figure you can see a sequence with repeats.

Figure: The dot plot of a sequence showing repeated elements.

Frame Shifts

Frame shifts in a nucleotide sequence can occur due to insertions, deletions or mutations. Such frame shifts can be visualized in a dot plot as seen in figure. In this figure, three frame shifts for the sequence on the y-axis are found.

- Deletion of nucleotides

- Insertion of nucleotides

- Mutation (out of frame)

Figure: This dot plot show various frame shifts in the sequence.

Sequence Inversions

In dot plots you can see an inversion of sequence as contrary diagonal to the diagonal showing similarity. In figure you can see a dot plot (window length is 3) with an inversion.

Figure: The dot plot showing an inversion in a sequence.

Low-complexity Regions

Low-complexity regions in sequences can be found as regions around the diagonal all obtaining a high score. Low complexity regions are calculated from the redundancy of amino acids within a limited region. These are most often seen as short regions of only a few different amino acids. In the middle is a square shows the low-complexity region of this sequence.

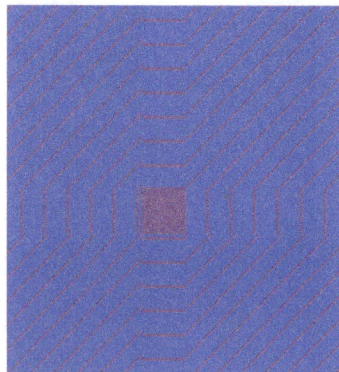

Figure: The dot plot showing a low-complexity region in the sequence. The sequence is artificial and low complexity regions do not always show as a square.

Structural Annotation

A DNA sequence, for instance a genome sequence, has much more value if we can annotate were are the different features like promoters, exons, introns, CDSs, transposons, etc. The annotation of those regions in a sequence is the structural annotation. This structural annotation is usually acompanied by a further functional annotation that will try to show the functions for these different regions. These annotations are very helpful for the users of the genomic sequence. Let's suppose that we link a disease to a region of a genome. If we have a poor annotation or we just have no annotation at all we won't know which genes are located close to that region or even if we have any gene at all. Having a good quality annotation allow us to make sense of the region under study.

We can base the structural annotation on experimental data, like ESTs, or we can do bioinformatic analyses in the sequence itself. These later annotations that start with just the sequence to be annotated are usually refered to as ab initio annotations.

There are different bioinformatic analyses that we can use to predict the gene structure (promoter, exons and introns), the alternative splicing, the coded protein, etc.

We have to take into account that we can have different degrees of confidence in these annotations. They can be supported by different amounts of experimental evidence and the methods used to create them can have different degrees of sensitivity and specificity.

Restriction Maps

A restriction map is a kind of physical map that shows the location of the restriction sites within a DNA sequence.

To create a restriction map we need a database with the restriction enzyme sites. One popular option is to use Rebase. Then a software programe can be used to look for the restriction sites in the nucleotide sequence. remap from EMBOSS is one of the options.

```
                            Cfr10I        Hin4I         PvuI
          Hpy8I              |     McrI    Hin4I         McrI
            \                 \     \        \            \
ACAAATCGTAAACAACTAAAAAATGTTTGCCGGTCGTTTGATGGTCCGTTCGATCGTTGG
        10        20        30        40        50        60
----:----|----:----|----:----|----:----|----:----|----:----|
TGTTTAGCATTTGTTGATTTTTTACAAACGGCCAGCAAACTACCAGGCAAGCTAGCAACC
        /                    //     /                 /
      Hpy8I                   ||    Hin4I            McrI
                             ||    Hin4I            PvuI
                            |Cfr10I
                            McrI
```

A restriction map can be used to map the sequences against a bigger physical map. Nowadays is quite common to use the maps created by the Bionano company for that porpose.

Translating a DNA Sequence

Predicting which peptide can be encoded by a DNA sequence is quite straightforward. We need the DNA sequence, the genetic code and the location of the start codon.

A DNA sequence is transcribed to RNA and translated to a peptide by using the genetic code.

```
DNA AGG TTT ACA TGT AGA GGA TGA

RNA AGG UUU ACA UGU AGA GGA UGA

PROT Arg Phe Thr Cys Arg Ala Fin
```

Although the RNA uses Uracil instead of Thymine it is very uncommon to have sequence files that correspond to RNA sequences with Us. The standard is to use in the sequence files T for both DNA and RNA.

To translate a nucleotide sequence we need to decide which genetic code to use. All known life uses a very similar genetic code. The genetic code is said to be universal, but it is in fact almost universal, there are slight changes in the code. For instance, the human mithocondria uses a slight variation of the genetic code.

A nucleotide sequence has six possible translations, three in the forward direction and six in the reverse and complementary strand.

```
Forward

    DNA:   AGGTTTACATGTAGAGGA

    1fr:   ArgPheThrCysArgGly

    2fr:    GlyLeuHisValGlu

    3fr:     ValTyrMetAm*Arg

Reverse and complementary

    DNA:   TCCTCTACATGTAAACCT

    1fr:   SerSerThrCysLysPro

    2fr:    ProLeuHisValAsn

    3fr:     LeuTyrMetOc*Thr
```

EMBOSS has two programs to get the six possible translations transeq and sixpack.

```
    T  N  R  K  Q  L  E  N  V  C  R  S  F  D  G  P  F  D  R  W    F1
    Q  I  V  N  N  *  K  M  F  A  G  R  L  M  V  R  S  I  V  G    F2
    K  S  *  T  T  K  K  C  L  P  V  V  *  W  S  V  R  S  L  V    F3
  1 ACAAATCGTAAACAACTAAAAAATGTTTGCCGGTCGTTTGATGGTCCGTTCGATCGTTGG 60
    ----:----|----:----|----:----|----:----|----:----|----:----|
  1 TGTTTAGCATTTGTTGATTTTTTACAAACGGCCAGCAAACTACCAGGCAAGCTAGCAACC 60
    V  F  R  L  C  S  F  F  T  Q  R  D  N  S  P  G  N  S  R  Q    F6
    X  L  D  Y  V  V  L  F  H  K  G  T  T  Q  H  D  T  R  D  N    F5
    C  I  T  F  L  *  F  I  N  A  P  R  K  I  T  R  E  I  T  P    F4


    S  G  M  L  G  H  H  G  Q  V  V  K  A  P  S  T  R  Q  P  S    F1
    R  A  C  L  A  T  M  G  R  W  S  K  P  Q  A  H  A  S  Q  V    F2
    G  H  A  W  P  P  W  A  G  G  Q  S  P  K  H  T  P  A  K  *    F3
 61 TCGGGCATGCTTGGCCACCATGGGCAGGTGGTCAAAGCCCCAAGCACACGCCAGCCAAGT 120
    ----:----|----:----|----:----|----:----|----:----|----:----|
 61 AGCCCGTACGAACCGGTGGTACCCGTCCACCAGTTTCGGGGTTCGTGTGCGGTCGGTTCA 120
    D  P  M  S  P  W  W  P  C  T  T  L  A  G  L  V  R  W  G  L    F6
    T  P  C  A  Q  G  G  H  A  P  P  *  L  G  L  C  V  G  A  L    F5
    R  A  H  K  A  V  M  P  L  H  D  F  G  W  A  C  A  L  W  T    F4


    D  P  A  Q  H  T  S  D  S  R  S  C  Y  S  M  R  G  I  H  C    F1
    I  L  P  S  T  P  A  I  A  A  V  A  I  Q  C  E  E  F  T  A    F2
    S  C  P  A  H  Q  R  *  P  Q  L  L  F  N  A  R  N  S  L  P    F3
121 GATCCTGCCCAGCACACCAGCGATAGCCGCAGTTGCTATTCAATGCGAGGAATTCACTGC 180
    ----:----|----:----|----:----|----:----|----:----|----:----|
121 CTAGGACGGGTCGTGTGGTCGCTATCGGCGTCAACGATAAGTTACGCTCCTTAAGTGACG 180
    S  G  A  W  C  V  L  S  L  R  L  Q  *  E  I  R  P  I  *  Q    F6
    H  D  Q  G  A  C  W  R  Y  G  C  N  S  N  L  A  L  F  E  S    F5
    I  R  G  L  V  G  A  I  A  A  T  A  I  *  H  S  S  N  V  A    F4
```

Backtranslation

We can also backtranslate a peptide sequence to a nucleotide sequence. In this case, due the degeneracy of the genetic code we will get many nucleotide uncertanities.

```
> peptide

    ACDEFGHIKLMNPQRSTVWY*

    >reverse translation

    gcntgygaygarttyggncayathaarytnatgaayccncarmgnwsnacngtntggtaytrr
```

You can backtranslate a peptide sequence in EMBOSS with backtranseq. There is also the possibility of infering the most probable codons for the backtranslation by using a codon usage table. EMBOSS can do it with the program backtranambig.

Codon Usage Tables

There is a codon usage bias, not all synonymous codons are used in the same frequency. Different organisms have different biases. They tend to prefer one codon over others to code for a particular aminoacid.

By counting how many times an aminoacid is used in the proteins coded by an organism nucleotide sequences a Codon Usage Table can be compiled for every species.

```
Drosophila melanogaster [gbinv]: 42417 CDS's (21945319 codons) fields: [triplet] [amino acid] [fraction] [frequency: per thousand] ([number])

UUU F 0.38 13.2 (289916)   UCU S 0.08  7.0 (154186)   UAU Y 0.37 10.8 (236811)   UGU C 0.29  5.4 (118088)
UUC F 0.62 21.8 (479372)   UCC S 0.24 19.6 (429341)   UAC Y 0.63 18.4 (403675)   UGC C 0.71 13.2 (288853)
UUA L 0.05  4.5 ( 97715)   UCA S 0.09  7.8 (171695)   UAA * 0.41  0.8 ( 17807)   UGA * 0.25  0.5 ( 10767)
UUG L 0.18 16.1 (353621)   UCG S 0.20 16.6 (365159)   UAG * 0.33  0.7 ( 14362)   UGG W 1.00  9.9 (217518)

CUU L 0.10  9.0 (196787)   CCU P 0.13  6.9 (151856)   CAU H 0.40 10.8 (236061)   CGU R 0.16  8.8 (192276)
CUC L 0.15 13.8 (303153)   CCC P 0.33 18.1 (396168)   CAC H 0.60 16.2 (354699)   CGC R 0.33 18.0 (395106)
CUA L 0.09  8.2 (180360)   CCA P 0.25 13.5 (297071)   CAA Q 0.30 15.6 (342415)   CGA R 0.15  8.4 (185119)
CUG L 0.43 38.2 (839127)   CCG P 0.29 15.8 (347206)   CAG Q 0.70 36.1 (792657)   CGG R 0.15  8.2 (180473)

AUU I 0.34 16.6 (363497)   ACU T 0.17  9.5 (208889)   AAU N 0.44 21.0 (460669)   AGU S 0.14 11.5 (252555)
AUC I 0.47 22.9 (502821)   ACC T 0.38 21.3 (467509)   AAC N 0.56 26.2 (575297)   AGC S 0.25 20.4 (447808)
AUA I 0.19  9.5 (208315)   ACA T 0.20 11.0 (241893)   AAA K 0.30 17.0 (372524)   AGA R 0.09  5.1 (112784)
AUG M 1.00 23.6 (518200)   ACG T 0.26 14.4 (315479)   AAG K 0.70 39.5 (866960)   AGG R 0.11  6.3 (137902)

GUU V 0.19 11.0 (240735)   GCU A 0.19 14.4 (315879)   GAU D 0.53 27.6 (604730)   GGU G 0.21 13.3 (291161)
GUC V 0.24 13.9 (304893)   GCC A 0.45 33.6 (736394)   GAC D 0.47 24.6 (540386)   GGC G 0.43 26.7 (587016)
GUA V 0.11  6.4 (139476)   GCA A 0.17 12.8 (280181)   GAA E 0.33 21.1 (462468)   GGA G 0.29 18.0 (395377)
GUG V 0.47 27.8 (609794)   GCG A 0.19 14.0 (307977)   GAG E 0.67 42.5 (933622)   GGG G 0.07  4.7 (102708)
```

ORFs

An Open Reading Frame (ORF) is the strecth of a reading frame that do not contain a stop codon. An AUG codon in a ORF might indicate the translation start.

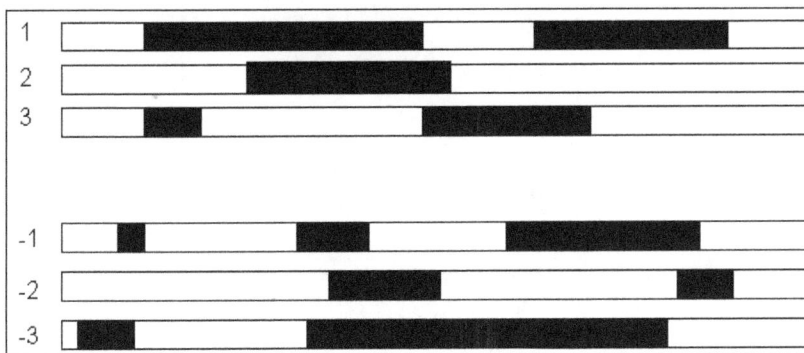

There is software in EMBOSS to show the ORFs (plotorf) and to get their translations (showorf).

CDS Prediction

Long ORFs might indicate the presence of a coding sequence and of a corresponding gene. In the procaryotes it is easier to detect long ORFs because they are not broken by the introns. In the eucaryotes the ORFs will be fragmented in multiple regions and they will be harder to detect. To be able to detect the CDSs in th eeucaryotes we need also to infer the gene structure, the location of exons and introns and the splicing variants.

A nucleotide sequence can have several open reading frames. There are several confusing factors:

- The start codon might be outside the sequence.

- There might be sequencing errors that create artifactual stop codong or that shift the frame.

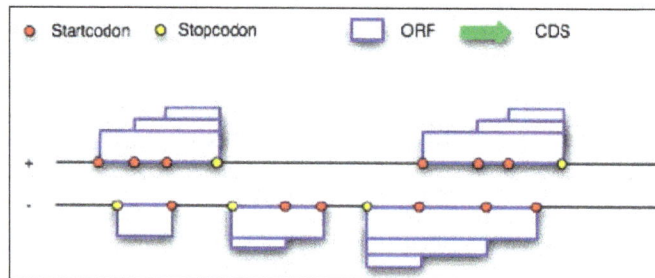

There are different hints that can point to the ORF that corresponds to the real CDS.

The codon use table for the species can be used to detect the real CDS. The real CDS will tend to use the codond prefered by the species, while the other ORFs will use codons at random. From these fact the coding likelihood of all six frames of a nucleotide sequence can be calculated.

Even if we don't have the codon use table for the species we can still use the codon bias to help with the detection of the CDSs. The codon bias will create a bias in the nucleotide frequencies of the third base. We can plot the GC content of the third base of the codons along a window for all six possible frames and detect biases there.

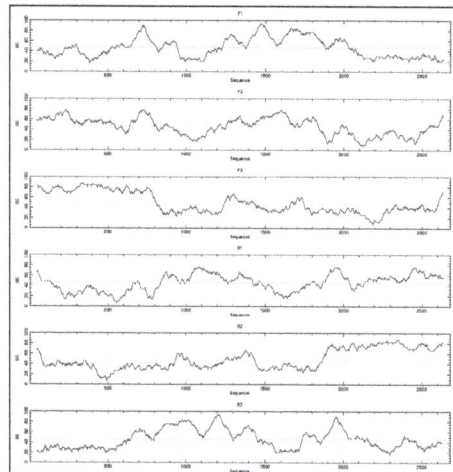

Another hint that can be used to detect CDSs is the protein conservation. Most proteins have similar proteins from other organisms in the database. BLAST can translate the six frames and can do a search aginst the known proteins. A similarity in a fragment of a frame with a known protein is an indication that that fragment is part of a real CDS.

Sequencing erros might produce frame shifts. Also, in the eucaryotes, the CDS will be split by the introns and can be divided in different jump. There are software that integrates the codon usage bias, the presence of start and stop codons and the sequences required by the splice sites to calculate a probability of any region of a nucleotide sequence being a coding region. One example of such software is GeneMark.

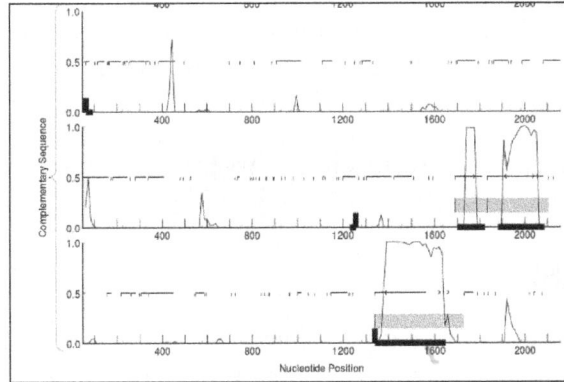

Promoter Prediction

It would be very valuable to locate the promoter union sites for the standard proteins of the transcription process and the union sites for the transcription factors. Locating a promoter region in a genomic sequence is not a straightforward task. Transcription factor targets have several characteristics that difficult their location, they are:

- Small

- Not very conserved.

Small sequences can be found just by chance locating such a sequence is not a guarantee that we have found a real transcription factor union site. Also their variability might let us think that there are many union sites, but most of them won't be real. Just by looking at those small sequences will create a lot of false positives.

Another factor that has been demonstrated to influence the gene expression is the histone location. A DNA region heavily coiled around the histones will have less chances to get accessed by the proteins involved in the transcription. Histone location can be influence by the DNA composition, by the methilation or by other factors that do not depend of the DNA sequences, so it is difficult to predict it.

Typical Promoter Sequences

There are sequences that are typically found in most promoters. Their composition depend on the species.

In bacteria there is a Pribnow box. Its consensus sequence is TATAAT and it is located at -10 upstream of the transcription starting site. A similar sequence to this one will be found in most promoters, but the sequence might be slighty different in some of them.

The Pribnow box has a similar function to the TATA box found in eucaryotes and archea, it is recognized by the RNA polimerase during the initiation of the transcription.

There is also a TTGACA usually found at -35 in bacteria and other less conserved regions.

In eucaryotes and archea the TATA box is located usually between -20 or -35 relative to the transcription starting site. The TATA box has a core conserved core of TATAAA which is usually followed by three or more adenines.

There are also in eucaryotes other sequences found that are not as conserved, like the CCAAT box and the GC box. The GC box is usually located 110 bases upstream of the transcription starting site and its consensus sequence is GGGCGG.

The CCAAT box is usually located 60 to 100 bases upstream of the transcription starting site and its consensus sequence is GGCCAATCT.

Transcription Factor Sites

There are many DNA motifs recognized by the transcript factors. Each transcription factor recognizes some particular sequences different from the other ones. These sequences are usually small and the transcription factor, in most cases, can lock to them even if they chage a little, so they can ve more or less conserved between different promoters.

Every gene is controlled by different transcription factors and the location of the sequences recognized by the factors can be quite variable and can be far appart from the transcrition initiation site.

Describing the Sites

The sites which are binded by the transcription factors are variable, but we have to describe them. We can use different ways to describe those sites.

We can use a consensus sequence. A consensus sequence shows the most abundant nucleotide for each position and ignores the rest of the variation. For instance, TATAAT is the consensus sequence for the Pribnow box, but that does not mean that all the Pribnow boxes have these sequence.

Sequence patterns. We can show all possible nucleotides for each position. Like TATAAW or TATAA(A/T). These patterns will match with all Pribnow boxes. The problem with them is that they do not include information about the sequence. Maybe one of the alternative nucleotides is found many more times the others. It could be well the case that TATAAA is more common than TATAAT, but that information is lost in the sequence pattern.

Position-specific scoring matrix. They include the frequency of all the residues for every position.

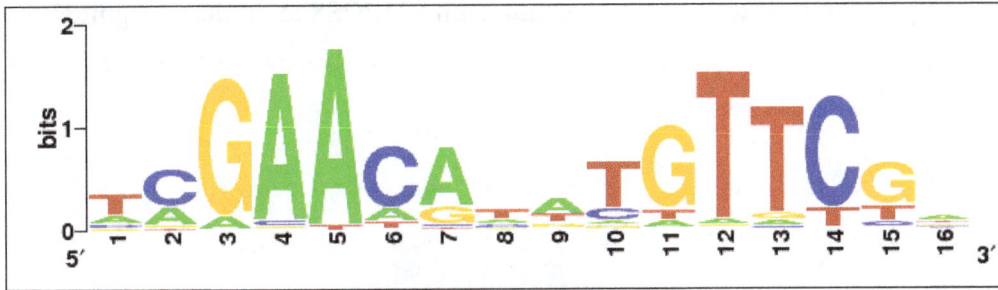

Detecting Transcription Factor Sites

There are different computational methods to predict the biding sites for the transcription factors, but none of them can do highly accurate predictions. The best of them are capable of recognizing at most 60% of the promoters, and they product many false positives. There are several programs to do the transcription site an promoter detection like: tfscan, jaspscan or GrailEXP.

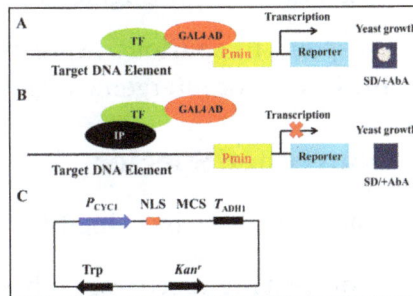

We can also align the promoter regions from very close species. The sites binded by the transcription factors will tend to be more consereved than the rest of the promoter because they have more functional relevance.

Gene Structure Prediction

The gene structure prediction is the process of predicting the location of the genes in a genomic sequence as well as its elements like promoters, exons, introns and coding regions. These genes can be protein coding or RNA coding.

CDSs and ORFs

In procaryotes the problem of looking for protein coding genes is easier because the CDSs won't be fragmented in pieces by the introns and the appearance of long ORFs will be a strong indication of their presence. In eucaryotes the problem is harder to solve because the introns split the CDSs and the ORFs won't be as informative. Also, the size of the non-coding regions can be quite large, specially the introns. In locating genes that do not code for proteins we won't be able to use any ORF hints.

Alignment with ESTs or cDNAs

We can use the information provided by the ESTs or cDNA to locate regions expressed in the genome. These regions will tend to be genes. We can also use this information to build the structure of the different mRNAs that a gene can produce.

We can use the programs est2genome and esim4 from EMBOSS to do these alignments.

```
Note Best alignment is between forward est and forward genome, and splice sites imply forward gene
Exon        340 100.0 1200 1539              1   340
+Intron     -20   0.0 1540 1608
Exon        544  98.6 1609 2171            341   900

Span        864  99.1 1200 2171              1   900

Segment     340 100.0 1200 1539              1   340
Segment     366 100.0 1609 1974            341   706
Segment     184  97.4 1978 2171            707   900
```

To be able to use these method of detecting genes we need to have transcriptomic sequence data. Even with these data at hand we should consider different limitations:

- We can have low expressed genes that are not represented in our transcriptomic data.

- We could have some mRNAs not completely covered.

- We have to infer the mRNA structure from the fragmented ESTs and that is not a trivial task.

- We can have ESTs that correspond to spurious expression like non-mature mRNAs or transposable expression in the introns or intergenic regions.

Comparison with Proteins from other Species

We can compare our genome to a protein database by using BLAST.

If we detect a match in a genome to a protein sequence from other species we have a hint that that genomic sequence might be part of a gene. Of course these method has also some limitations:

- There might be no proteins similar to some of the genes in our species.

- We can have pseudogenes that could match with some protein despite not being real genes.

Combining Different Evidences

There are different programs that try to predict the gene structure by taking into account the different evidences. They take into account:

- The codon usage.

- Splicing sites.

- ORFs.

- The GC content. Gene regions tend to have a higher GC content.

- EST and cDNA alignment.

- Comparisons with sequence database.

Different programs will create different predictions even when the same evidence is considered. So the prediction will depend on the program and the evidence used. Of course, the more evidence and the higher quality that evidence is the easier will be to infer the real gene structure.

Role of Statistical Methods in Bioinformatics

Data analysis is seen as the largest and possibly the most important area of microarray bioinformatics. Statistical analyses for differentially expressed genes are best carried out via hypothesis tests rather than using a simple fold ratio threshold. More complex data may require analysis via ANOVA or general linear models and may be also include bootstrapping. Principal Component Analysis (PCA) and Multidimensional Scaling (MDS) provide a good way to visualise data without imposing any hierarchy on them. Hierarchical clustering can be used to identify related genes or samples and portray the usage of dendrogram. There are several methods for classifying samples, each with advantages and disadvantages, including: K-nearest neighbour, centroid classification, linear discriminant analysis, neural network, support vector machines.

It was well understood that computing would play a vital role in the future progress of statistics. Access to elaborate algorithms on computers increased the awareness of more recent methodological developments in statistics. According to the definition proposed by A. Westlake: "Computational statistics is related to the advance of statistical theory and methods through the use of computational methods. This includes both the use of computation to explore the impact of theories and methods, and development of algorithms to make these ideas available to users." Computation in statistics is based on algorithms which originate in numerical mathematics or in computer science. The core topics of numerical mathematics are numerical linear algebra and optimization techniques but practically all areas of modern numerical analysis may be useful. The group of algorithms highly relevant for computational statistics from computer science is machine learning, artificial intelligence (AI), and knowledge discovery in data bases or data mining. These developments have given rise to a new research area on the borderline between statistics and computer science. Besides the difficulties resulting from new problems in various research areas, for example analysis of microarrays in biology, the following three interwoven challenges for computational statistics: handling of problems stemming from new data capture techniques, from the complexity of data structures, and from the size of data. The summary of the different subjects of science interrelationship.

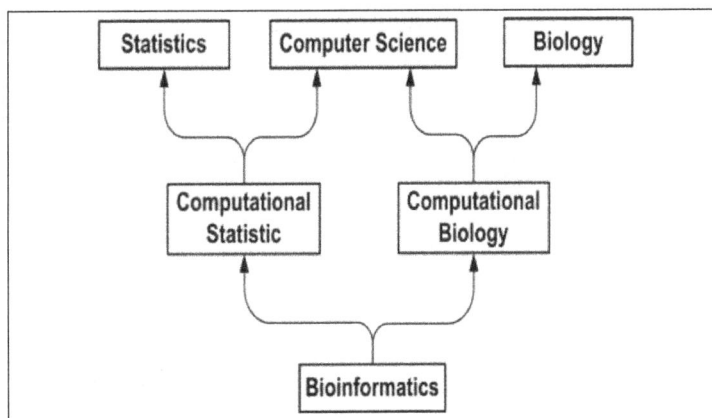

Figure: The interrelationship of the different subjects of sciences

Bioinformatics has grown into a large topic, but still one of the most widely used tools in bioinformatics is that for searching a sequence database for all sequences similar to a given query sequence. There are three main bioinformatics problems:

- Connection with "Dogma": sequence, structure and function.

- Connection with data: keeping, access and analysis.

- Biological process simulation: protein structure, (molecular dynamics), biological networks.

Data Mining Techniques and Statistical Methods Comparison

A variety of techniques have been developed over the years to explore for and extract information from large data sets. At the end of the 1980s a new discipline, named data mining, emerged. Traditional data analysis techniques often fail to process large amounts of data efficiently. Data mining is the process of discovering valid, previously unknown, and ultimately comprehensible information from large stores of data. Data Mining is the process of extracting knowledge hidden in large volumes of raw data. Data mining automates the process of finding relationships and patterns in raw data and delivers results that can be either utilized in an automated decision support system or assessed by a human analyst. Modern computer data mining systems self-learn from the previous history of the investigated system, formulating and testing hypotheses about the rules which this system obeys. In general, data mining techniques can be divided into two broad categories: predictive data mining and discovery data mining.

Predictive data mining is applied to a range of techniques that find relationship between a specific variable (called target variable) and the other variables in your data. The following are examples of predictive mining techniques:

- Classification is about assigning data records into pre-defined categories. In this case the target variable is the category and the techniques discover the relationship between the other variables and the category.

- Regression is about predicting the value of a continuous variable from the other variables in a data record. The most familiar value prediction techniques include linear and polynomial regression. Discovery data mining is applied to range of techniques that find patterns inside your data without any prior knowledge of what patterns exist. The following are examples of discovery mining techniques:

  ○ Clustering is the term for range of techniques, which attempts to group data records on the basis of how similar they are.

  ○ Association and sequence analysis describes a family of techniques that determines relationship between data records.

The particularity of contemporary requirements for the data processing is the following: the data have the unlimited quantity, the data differ (quantitative, qualitative and textual), the results should be particular and comprehensible and the tools for the processing of raw data should be easy to use. The modern technology of Data Mining (discovery-driven data mining) is based on the concept of patterns, reflecting fragments of polydimensional relationships within the data. These patterns are regularities, which are characteristic to the data sub-retrievals that can be reflected compactly in the form, which is easy to comprehend for the human. The search for the patterns is carried out by means of techniques, which are not limited by a priori proposals about the structure of retrieval and type of the value distribution of indicators to be analyzed.

The traditional mathematical statistics, for a long time applying for the role of the basic tool of data analysis, clearly gave up in front of the problems coming into existence. The main reason – the concept of the mean of retrieval that leads to the operations on the fictitious values. The techniques of mathematical statistics proved to be useful mainly for the verification of preliminary defined hypotheses (verification-driven data mining) and for the 'rough' research analysis that forms the basis of operative analytical data processing (online analytical processing, OLAP). At the same time the strong correlation exists between data mining and statistical methods, because statistical methods can be divided into the similar categories as data mining techniques: dependence methods and interdependence methods. The objective of the dependence methods is to determine whether the set of independent variables affects the set of dependent variables individually and/or jointly. That is, statistical techniques only test for the presence or absence of relationships between the two sets of variables. At the same time there exist such data sets for which it is impossible to designate conceptually the set of variables as dependent or independent. For these types of data sets the objectives are to identify how and why the variables are related among themselves. Statistical methods for analyzing these types of data sets are called interdependence methods. The classification of the data mining and statistical methods is the following:

- Data Mining:
  - Predictive techniques: Classification, Regression.
  - Discovery techniques: Association Analysis, Sequence Analysis.
- Clustering.
- Statistical methods:
  - Dependence methods: Discriminant analysis, Logistic regression.
  - Interdependence methods: Correlation analysis, Correspondence analysis, Cluster analysis.

Introduction to Random Variables and Probability Distributions

Firstly, we need to make some informal definitions for key phrases which will be used liberally throughout this course.

- Random experiment - Experiments for which the outcome cannot be predicted with certainty.

- Random variable - The outcome of a random experiment. Conventionally written with uppercase symbols e.g.: X,Y, etc.

- Discrete random variable - A numerical quantity that randomly assumes a value drawn from a finite set of possible outcomes. For example, the outcome of a dice throw is a discrete random variable with a solution space: {1,2,3,4,5,6}.

- Continuous random variable - Similar to the discrete case, but this time the solution space consist of a range of possible values, with (in principle) infinite resolution.

- Probability distribution - This is a function, PX(x) over the solution space of the random variable, yielding the probability of occurence for each potential outcome. Again this can be differentiated into discrete and continuous instances. Probability distributions are constrained by the following condition:

$$\int_x P_X(x)\,dx = 1.$$

For a discrete random variable X, the probability distribution is often represented in the form of a table containing all possible values which the variable can take, accompanied by the corresponding probabilities. In the conventional view, these are interpreted as the relative frequencies of occurence of the various values. An alternative perspective known as the Bayesian framework, in which the probabilities are treated as subjective measures of belief in certain outcomes of the random experiment.

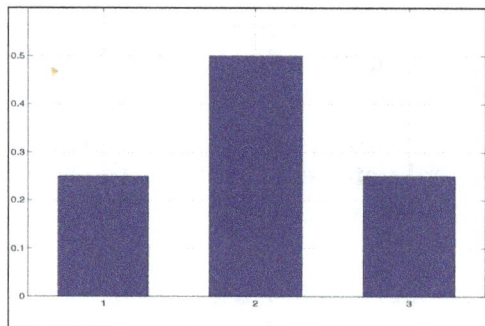

Figure: Probability distribution for coin toss experiment

A good example is the result of a coin toss experiment, which is performed by tossing a fair coin twice, and recording the number of heads observed. Assuming this experiment is performed a large number of times, we can expect that the results will occur approximately according to the following frequencies:

Number of heads (X)	Relative frequency
0	0.25
1	0.5
2	0.25

Clearly, the most straightforward way in which the probability distribution may be obtained is by repeating an experiment a large number of times, then compiling and tabulating the results. These may then be presented as a histogram depicting the probabilities of each of the outcomes. For our experiment above, an idealised graph.

Probability Distribution Functions

A probability distribution function or PDF is simply a function defined over the entire solution space (i.e. the space of all possible values which the random variable is able to return) which allows the probability or probability density at each potential solution to be determined analytically.

Many such functions have been proposed, corresponding to a variety of theorised situations. However, in real life experimental conditions such conditions are rarely achieved exactly, which means that the actual distributions from which real-life data is sampled often deviate from these idealised distribution functions. Nevertheless, for practical reasons and mathematical tractability, it is the accepted practice to model real life distributions by fitting one of the existing classes of distribution functions to match the data.

One Bernoulli Trial

A Bernoulli trial is a single trial with two possible outcomes, often called "success" and "failure". The probability of success is denoted by p and the probabilty of failure, q, is simply given by 1-p, since there are no other possible outcomes of the experiment.

Hence, if we label a "success" as a 1 and a "failure" as a 0, we obtain the following formula for the outcome of a bernoulli trial:

$$P_X(x) = p^x (x-1)^{1-x}, .\ x = \{0,1\}$$

The Binomial Distribution

A Binomial random variable is the number of successes obtained after repeating a given Bernoulli trial n number of times, where the probabilities p and q are fixed for the duration of the experiment. There is also the added condition that the outcome of the successive Bernoulli trials be independent of one another.

For a Binomial random variable X, the probability distribution P_X is given by:

$$P_X(x) = {}^nC_r p^x (1-p)^{n-x}, x = 1,2...,n$$

where ${}^nC_r$ is the combination operator, which gives the number of ways in which you can select r items from a collection of n. Note that the distribution is described by two parameters, n and p, which together determine the characteristics of the resulting distribution function.

In the case where n = 20 and p = 0.5, the resulting binomial distribution.

The Poisson Distribution

One commonly encountered scenario is where the event of interest occurs a finite number of times within a given time interval.

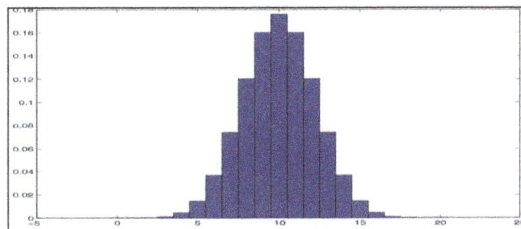

Figure: Binomial distribution with n = 20 and p = 0.5

Commonly quoted examples are the number of car accidents during a fixed period, the number of phone calls received, and so on. In such cases, while there is an infinitesimally small probability of the event occurring at a particular time instant, the actual number of time "instants" is extremely large (as the time duration is continuous, the number of instants is effectively infinite). Hence, it is often sufficient just to know the mean number of occurrences in a fixed time interval. The probability distribution for the number of occurrences can then be well approximated by the Poisson distribution, given by the following function:

$$f(x|\lambda) = \begin{cases} \dfrac{e^{-\lambda}\lambda^x}{x!} & \text{for } x = 0,1,2,... \\ 0 & \text{otherwise.} \end{cases}$$

Where λ is the mean number of occurrences in the time period of interest, and x is the actual number of occurrences. Clearly in this expression, e^λ serves as a normalising factor (since it does not depend on x), and the value of the probability is determined by the expression $\dfrac{\lambda^x}{x!}$.

The Uniform Distribution

Perhaps the most straightforward distribution function is the uniform distribution. A random variable is said to have a uniform distribution if the density function is constant over a given range (and zero elsewhere), i.e. all possible values within the accepted range of values have equal probability. For the range $a \geq x \geq b$, this is expressed as:

$$P(x) = \begin{cases} \dfrac{1}{a-b} & \text{for } a \geq x \geq b \\ 0 & \text{otherwise.} \end{cases}$$

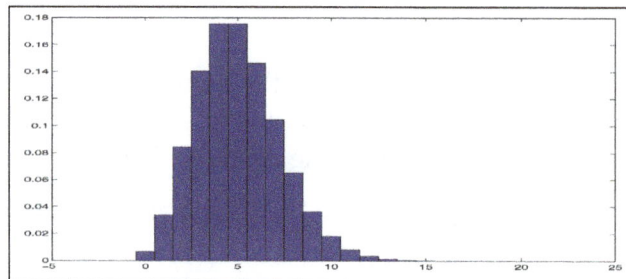

Figure: Poisson distribution with $\lambda = 5$

The Normal Distribution

By far the most common, and certainly the most important probability distribution that we will be studying is the normal or Gaussian distribution, shown in figure. A continuous random variable X drawn from a normal distribution has a density function:

$$P_X(x) = \frac{1}{\sqrt{2\pi}\sigma} e^{-\frac{(x-\mu)^2}{2\sigma^2}}$$

The density function is parameterised by the quantities σ and μ which represent the standard deviation and mean of the distribution respectively.

For a number of reasons, both theoretical and practical, the gaussian is the distribution of choice for many applications. However, two factors in particular account for its pre-eminence:

1. The mathematical properties of the density function for the normal distribution make it extremely easy to work with. The mean and variance of the distribution are immediately evident from the function - as a matter of fact, a gaussian is completely described by the mean and variance. Higher order cumulants of the distribution are zero.

2. Many naturally occurring phemomena often have distributions that are approximately normal. This is a direct consequence of the central limit theorem which states that the composite distribution resulting from the combination of a large number of independent random variables will converge to a normal distribution.

Characteristics of a Random Variable

The distribution of a random variable X contains all the information regarding the stochastic properties of X. However, in many cases, it is difficult to represent this information as the PDFs of real random variables can often be complex and not easily characterised by one of the existing families of analytical distributions. In this section we study ways by which a distribution may be "summarised" to give a general idea of its key properties without having to describe the entire distribution.

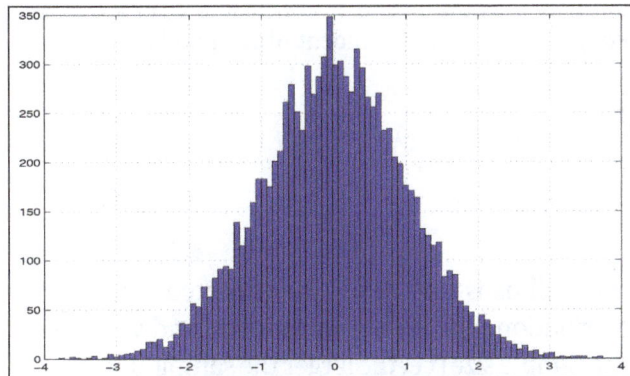

Figure: Histograms for samples drawn from a zero mean, unit variance normal distribution

Expectation

One of the most commonly invoked quantities is the expectation of a random variable. Denoted by E[x], this is defined as:

$$EX(x) = \int_{x=-\infty}^{\infty} x . P_X(x) \, dx$$

The number E[x] is also called the expected value of X or often simply the mean of X. Note that it is analogous to the centre of gravity of a physical object. Effectively, the mean is the centre of mass of the probability density function.

From this discussion it is evident that the mean should in fact be distinguished from the arithmetic average of a sample of points, also known as the sample mean.

For a sample size n, this is:

$$X_n = \frac{1}{n}(X_1, X_2, ..., X_n).$$

The mean is related to the actual underlying distribution from which the data is sampled, whereas the average is a statistical measure that is derived from the samples themselves. In fact, it can be proven that if a collection of populations were drawn from a given distribution, the averages of the individual populations would themselves be distributed according to a normal distribution, with a mean and variance determined by the number of points in the samples. In fact, the exact values for these parameters may be easily determined thus:

$$E\left[\overline{X}_n\right] = E\left[\frac{1}{n}(x_1 + x_2 + ... + x_n)\right]$$
$$= \frac{1}{n}.nE[X_i]$$
$$= \mu.$$

The variance of the sample means can be predicted in a similar fashion thus:

$$Var\left(\overline{X}_n\right) = \frac{1}{n^2}Var\left(\sum_{i=1}^{n} x_i\right)$$
$$= \frac{1}{n^2}\sum_{i=1}^{n} Var(X_i) \,(x's \text{ are independent of one another})$$
$$= \frac{1}{n^2}.n\sigma^2$$
$$= \frac{\sigma^2}{n}.$$

What these these two results tell us is that, while the expected value of the sample mean will be the mean of the underlying distribution, this is only an estimate and varies with a given variance which is inversely proportional to the sample size (i.e. the larger the sample size, the more accurate the estimate).

Moments of a Distribution

The mean and variance are special cases of the moments of a probability distribution.

For a random variable X, the rth moment Mr (where r is any positive integer, is defined as:

$$M_r = E_X\left[x^r\right]$$
$$= \int_{-\infty}^{\infty} x^r Px(x)d.$$

It is also cannot be assumed that a certain moment of a given distribution exists. If a distribution is bounded (i.e.: if the PDF integrates out to one), then it is necessarily true that all moments exist. However, while it is possible for all moments to exist even if the PDF is not bounded, this is not necessarily true. It can be shown that if the rth moment of X exists, then all moments of lower order must also exist.

Moment Generating Functions

Given the density function, how can find the moments of a distribution? In many cases, this can be obtained directly but often it can be quite challenging. One approach by which a given moment may sometimes be conveniently calculated is via a moment generating function.

Distribution Functions of more than One Random Variable

It is possible to combine PDFs from separate random variables to form composite distributions. In such cases, it is useful to be able to classify these according to their respective functions. These help to clarify what a distribution function says about a pair (or more) of random variable. In particular, we identify three common classes into which composite PDFs may fall.

Joint Distributions

For this, and all proceeding examples in this section, we will concentrate on the case where there are two random variables, X and Y, which are not necessarily independent. All examples can easily be generalised to the case of multiple random variables.

Consider the case where we sample simultaneously from X and Y, i.e. we conduct a joint experiment. What is the probability of observing a particular pair of outcomes? In this case, we can formulate the answer as a new composite distribution function which extends over the combination of the solution spaces of the two random variables.

To help visualise this, let us assume that X and Y are two discrete random variables with the solution space defined by X, Y $\in$ {1, 2, 3, 4, 5}. In this case, the possible combinations of values which the joint random variable (X, Y) can assume are shown in figure. For each of the points in the grid, we can now assign a probability of the corresponding outcome of the joint experiment. These probability values are denoted by P(X, Y), and are given by the joint probability distribution of X and Y.

Conditional Distributions

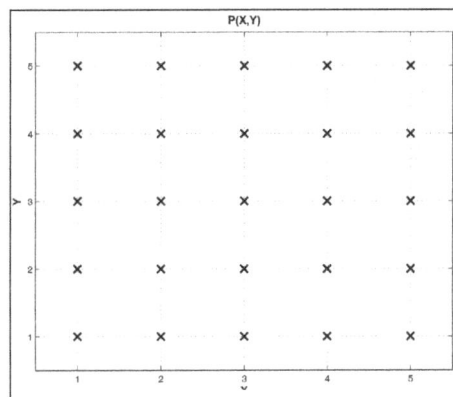

Figure: Possible combinations of values for discrete random variables X and Y

Suppose that we already know the outcome of experiment Y. Clearly this would greatly limit the number of possible outcomes in the joint solution space. In our current example, since we are only dealing with two variables, this effectively reduces the solution space to one dimensional. It is clear that, for any given value of Y, the corresponding probability for a particular value of X can be obtained simply by reading along the particular row corresponding to the incident value of Y.

We call this new distribution the conditional distribution of X given Y. Equivalently, it is normal to speak of the probability of X conditional upon a certain value of Y. Mathematically, this is written as P(X|Y), and is derived from:

$$P(X|Y) = \frac{P(X,Y)}{P(Y)}$$

Note how it can easily be seen from figure that this corresponds to the joint distribution values for the required values of X (along the row corresponding to the incident value of Y), normalised by the sum of all the joint probability values along the row.

Marginal Distributions

In the final example, consider the situation where we are not interested in the outcome for experiment Y, i.e.: we are only interested in the outcome of X. For a given $X_n = x$, we can obtain the unconditional probability by summing over P(X, Y) for all the possible values of Y. This is a process called marginalisation and is written as follows:

$$P(X) = \sum_Y P(X,Y)$$

The resulting distribution, P(X), is then called the marginal distribution.

Independent Random Variables

Before proceeding further, this is a suitable point for the introduction of the concept of statistical independence when applied to random variables. In many cases, "independence" as used in statistics corresponds well with the general meaning of the word, as used in everyday situations. i.e., a given random experiment is independent of another random experiment if the associated random variables do not depend on each other in any way. For example, the result of two successive coin-tosses occur completely randomly and are independent of one another.

However, it is still useful for a formal definition be given. We say that two random variables X and X are considered statistically independent if, and only if, the joint distribution of the two variables is equal to the product of the two marginal distributions, written as:

$$P(X,Y) = P(X)P(Y)$$

In such a case, the grid in figure becomes a multiplication matrix - where the values associated with the vertices can be found from the product of the unconditional probabilities $P(X)$ and $P(Y)$.

Estimation Theory

So far, we have covered some of the basic concepts of probability which provide the basis upon which the study of statistics is built. In particular, we would like to consider real world data as observations of some underlying generator. As was mentioned earlier in the notes, it is almost always impossible to study this underlying generator directly. However, what is commonly possible is to learn about its properties based on indirect observations.

We have seen that one way in which we can reason about this underlying probability distribution in a sensible way is if we assume some parametric distribution for it. For example, if we want to learn about the distribution of heights in the population of Malaysia, we can assume that it is drawn from a gaussian distribution (and in fact, it does, approximately!). The process by which we learn about the mean and variance of this distribution is a crucial activity in statistics and is widely referred to as estimation. As a loose guide, an estimator is some function or algorithm by which the realisations of a random variable are mapped to an estimate of the parameters of the underlying generator. Simply averaging a dataset provides a good example of an estimator that is very commonly used. It can be shown that the arithmetical average of a set of data provides an unbiased estimate of the expectation of the underlying distribution from which the data was drawn.

Maximum Likelihood Estimation

Broadly speaking, there are two approaches to statistics which, while actually sharing a lot of common ground, are widely regarded as being from opposing camps. One on hand, there is the "Frequentist" position, and on the other we have the Bayesian framework.

One popular method taken from the frequentist camp, is that maximum likelihood estimation. This is the procedure for estimating the parameters of the unknown model, by maximising the likelihood of the observed data. That is to say we would like to find:

$$\theta_{ML} = \arg\max_{\theta}\left[P(Y|\theta)\right]$$

Here, θ represents the parameters of the model which we would like to estimate, whereas Y denotes the available observations.

Example: Linear Regression

Suppose we have a set of paired values, x and y, which we assume are linearly correlated. Accordingly, we assume that the two are related by the expression y = Mx. Hence, we would like to estimate the value of the parameter M, which in this case is the gradient of the line obtained by plotting x vs y on an x – y plane. Finally, to obtain a maximum likelihood solution, we also need to assume some kind of noise model. This is necessary because real data is never exact - otherwise, we can obtain M simply by evaluating:

$$M = \frac{y_1 - y_2}{x_1 - x_2}$$

A commonly used assumption is that of gaussian noise. That is to say,

$$y = Mx + v$$

where $v \sim N(\mu, \sigma)$. Hence, the distribution of y conditional upon x is given by:

$$P(y \mid x) \sim N(Mx, \sigma)$$

$$\propto \exp\left[\left(\frac{y - Mx}{\sigma}\right)^2\right]$$

To simplify the maximisation of the likelihood, we now take the logarithm of the expression above. Note that this is acceptable since logarithm is a monotonic function - i.e.: it only increases in one direction, such that $\log(x_1) > \log(x_2)$ necessarily implies that $x_1 > x_2$. Taking the logarithm of $P(y|x)$ yields the log-likelihood term:

$$-\log P(y \| x) = -\left(\frac{y - Mx}{\sigma}\right)^2$$

We take the negative log likelihood - the reason for this will become clear shortly. We can now easily differentiate this with respect to M, and set to zero, to obtain:

$$\frac{d\left[-\log P(y \| x)\right]}{dm} = \frac{2}{\sigma^2}(y - Mx).x = 0$$

$$\therefore \quad x^T y = x^T x M$$

$$\therefore \quad M = \left(x^T x\right)^{-1} x^T y$$

The final left hand expression, $(x^T x) -1x^T$. This is called the pseudoinverse of x, and is commonly denoted as x. The evaluation of the pseudoinverse of a matrix is a common function which is widely available in statistics/mathematical packages - enabling maximum likelihood fitting of this sort to be performed with great ease. Nevertheless, it is useful and conceptually important to be aware of the underlying model - i.e. linear regression is actually equivalent to fitting a linear gaussian noise observation model to the data.

Bayesian Framework

The maximum likelihood method discussed above has proved very useful for many applications. However, it also has some shortcomings. In particular, by maximising over the parameter space, it is discarding all the possible model parameters in favour of one "optimal" solution. While this is a practical strategy in many instances it is also sensitive to the shape of the likelihood function. Take, for example, the likelihood function depicted in figure. This is an example of a bimodal distribution - in fact a mixture of two gaussian distributions. However, one of the density functions has a much smaller variance and as such is a lot more peaked. In fact however, the probability mass of the first distribution is only half that of the flatter distribution. This means that, while the maximum likelihood solution will be the peak of the first distribution, it is far more likelihood that the "optimal" parameters will lie somewhere in the region defined by the second distribution.

The Bayesian framework helps to overcome this problem by attempting to consider the entire PDF of the solution space, rather than just the mode. It is based on Bayes' theorem, which is given by:

$$P(X|Y) = \frac{P(Y|X)P(X)}{P(Y)}.$$

What this provides, in very general terms, is a means by which the conditional probability of a given model, X, given the available data, can be linked to the conditional probability of observing the data if the model were correct. In practice, the significance of this is that it gives a broad relationship for estimating the parameters θ of a proposed model provided based on observations derived from the true model. This is because it is much easier to estimate P(Y |X) than the other way around.

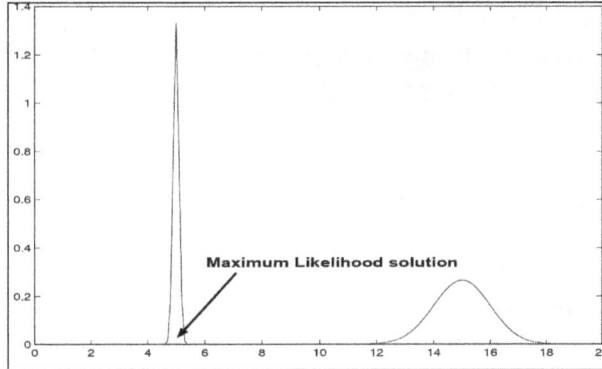

Figure : PDF consisting of a mixture of two gaussian distributions with $\mu_1 = 5$ and $\mu_2 = 15$ and $\sigma_1 = 0.1$ and $\sigma_2 = 1$ respectively

In Bayesian terminology, the terms in equation are often referred to as follows:

1. In common with frequentist terminology, P(Y|X) is called the likelihood function. This is basically the probability of observing the experimental data, given the proposed model,

2. P(X) is the prior distribution for the model. This terms allows any prior information regarding the model parameters to be incorporated into the inference. If no prior information is available, a suitable "initial guess" can be provided,

3. P(X|Y) is the posterior distribution. This reflects the knowledge we have regarding the model after having incorporated information contained in the observations,

4. Finally, the P(Y) in the denominator on the right hand side is called the evidence term. This is obtained by marginalising out X - and is thus constant for all values of X. Hence, its main function is as a normalising term.

The key concept in Bayesian analysis is that the distributions described above are constantly updated, to reflect the increase in our knowledge about the system being studied as new observations become available (it is also possible that very noisy observations will actually increase the degree of uncertainty by "spreading" the distributions). However, in general, the process of updating the distributions accompanies an increase in our knowledge of the system.

The general process of Bayesian learning is as follows:

1. Start by making an initial guess on the state of the system. This is the prior distribution P(X),

2. When we receive (or make) an observation, Y, we can calculate the likelihood P(Y|X) using the model X that we have assumed,

3. The combination of prior and likelihood can then be used to calculate the updated distribution for X, i.e. the posterior distribution P(X|Y),

4. The whole process can be repeated whenever a new observation (or set of observations) is obtained. However, at each step, the posterior of the previous step is used as the new prior distribution.

In this way, the Bayesian methodology also carries certain philosophical implications as well. In particular, it describes a systematic framework in which we may explicitly specify our belief in the parameters of a model, and a procedure through which this belief may be updated by comparison with observed data.

Markovian Dynamics

Hitherto, we have looked at probability distributions that do not change in time. The models that have been examined in the preceeding sections specify a static density function over the solution space, and it is assumed that observations may be made indefinitely without changing the probabilistic structure of the data.

In this section we introduce a modelling paradigm which allows for changes in the statistical properties of the data over time. Such dynamic models allow a much more general range of phenomena to be modelled.

Dynamical Processes

A dynamical process is basically one which changes over time. Essentially, these processes are regarded as being composed of an underlying "state" which evolves in time according to some dynamic evolution rule, often containing stochastic components. This is illustrated in figure, where xt represents the state of the system at time t, and yt the observation (also at time t).

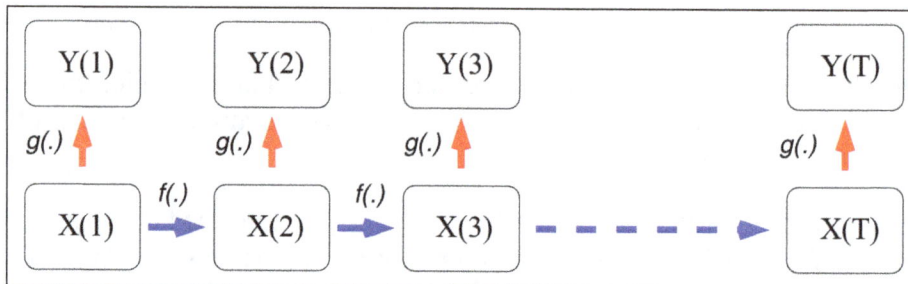

Figure: Block diagram depicting the evolution of a generic dynamical process through time

Markov Processes

The key elements in this model are the two functions f(.) and g(.). The function f(.) is known as the transition or evolution function and determines how the system changes over time. In general we would like to model cases where the following two relationships hold true:

$$x_t = f\left(x_{t-1}\right), \text{and}$$
$$y_t = g\left(yt\right)$$

$x_t = f(x_{t-1})$, is of particular significance as it indicates that the state of the system at time t is dependent only on the state of the system at time t − 1. This is known as the Markov property and any system in which this applies is a Markov process. Equation defines the relationship between the state of the system and the observations generated from it. Again, note that the observations at time t only depend on the state of the system at time t.

Biological Databases

A biological database is a collection of data that is organized so that its contents can easily be accessed, managed and updated.

There are two main functions of biological databases:

1. Make biological data available to scientists.

    As much as possible of a particular type of information should be available in one single place (book, site, database). Published data may be difficult to find or access, and collecting it from the literature is very time-consuming.

2. To make biological data available in computer-readable form.

    Since analysis of biological data almost always involves computers, having the data in computer-readable form (rather than printed on paper) is a necessary first step.

The computer became the storage medium of choice as soon as it was accessible to ordinary scientists. Databases were distributed on tape, and later on various kinds of disks. When universities and academic institutes were connected to the Internet or its precursors (national computer networks), it is easy to understand why it became the medium of choice. And it is even easier to see why the World Wide Web (WWW, based on the Internet protocol HTTP) since the beginning of the 1990s is the standard method of communication and access for nearly all biological databases.

As biology has increasingly turned into a data-rich science, the need for storing and communicating large datasets has grown tremendously. The obvious examples are the nucleotide sequences, the protein sequences, and the 3D structural data produced by X-ray crystallography and macromolecular NMR. An new field of science dealing with issues, challenges and new possibilities created by these databases has emerged: bioinformatics. Other types of data that are or will soon be available in databases are metabolic pathways, gene expression data (microarrays) and other types of data relating to biological function and processes.

One very important issue is the frequency and type of errors that the entries in a database have. Naturally, this depends strongly on the type of data, and whether the database is curated (modified by a defined group of people) or not. For the sequence databases, the errors may be either in the sequence itself (misprint, wrong on entry, genuine experimental error) or in the annotation (mistaken features, errors in references,). In the 3D structure database (PDB), structures have been deposited which were later discovered to contain severe errors. The error handling policy differs considerably between databases. If one needs to use any particular database heavily, then the implications of its particular policy need to be considered.

The data repositories more relevant to the biological sciences include:

- Nucleotide and protein sequences
- Protein structures
- Genomes
- Genetic expression

Main sequence databases:

- NCBI
- EMBL

Main protein databases:

- Uniprot
- PDB
- MMDB

Some genome databases:

- ENSEMBL (Human, mouse and others)
- SGD (Yeast)
- TAIR (Arabidopsis)

Human diseases:

- OMIM

Metabolic pathways:

- KEGG

Sequence Databases

A sequence database is a collection of DNA or protein sequences with some extra relevant information. The main sequence databases are Genbank and EMBL. Originally they were just sequence collections, but they have grown to store different biological databases heavily interconnected and they provide powerful interfaces to search and browse the stored information.

The sequences submitted to any of those databases are shared between them, so any sequence could be retrieved in the european or the american database. But they differ in the tools to search and browse the data and in some databases that provide extra information to the raw sequences like: mutations, coded proteins, bibliographical references, etc.

These databases are growing at an ever increasing fast pace. In June of 2007 there were 73 million sequences in Genbank and in August of 2015 there were 187 millions.

The sequences are split in these databases in different sections to ease the search. Among others, there are sections for mRNAs, publised nucleotide sequences, genomes, and genes.

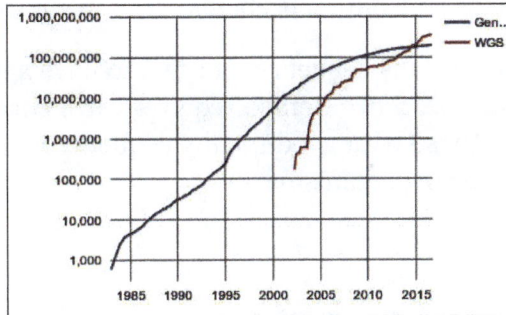

Genbank

Genbank is a public collection of annotated sequences hosted by the NCBI. Among other kinds of sequences Genbank includes messenger RNAs, genomic DNAs and ribosomic RNA.

Some characteristics:

- It is a public repository, any one can send sequences to it.

- There are sequences of different qualities, anything submitted is stored.

- There could be multiple sequences for the same gene or for the same mRNA.

- A sequence can have several versions that represent the modifications done by the authors.

Due to the huge amount of sequences stored to ease the search the databases are split in different divisions. These divisions follow two criteria: the species and type of sequence. Among the taxonomical divisions you can find: primate, rodent, other mammalian, invertebrate an others. The other divisions are related to the kind of sequences like: EST, WGS, HTGS, and many others. If you are looking for reads comming from the Next Generation Sequencing Technologies they are stored in a special division called SRA.

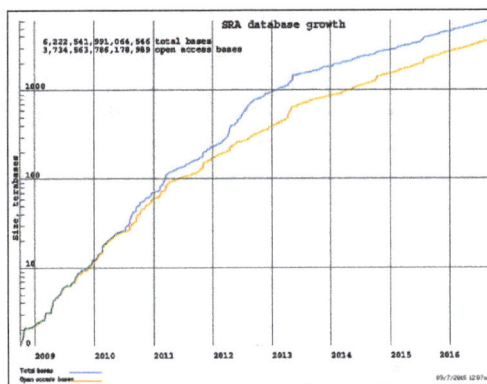

RefSeq

In RefSeq there are only well annotated and good quality sequences. It stores genomic, transcript and protein sequences and links the sequences that belong to a gene. It just has one representative sequence for each mRNA in a particular organism and, thus, it will have as many sequences as different transcripts and proteins coded for a particular gene in a particular organism.

It is not the aim of ReqSeq to have any sequence, but just to have a collection of well curated sequences. It is a secondary database. Since RefSeq requires extra curation work it is not available for all organisms, but only for those with good quality sequences. As of July of 2016 it has 65M proteins and 15M transcripts for 60K organisms.

UniProt

UniProt is a protein database that includes information divided in two sections: Swiss-Prot and TrEMBL. UniProt aims to store sequence and functional information for the proteins.

TrEMBL is automatically annotated while Swiss-Prot is reviewed manually by humans that add information by reviewing the literature. Due to this effort Swiss-Prot has information of a higher quality, but it has less sequences than TrEMBL.

UniProt also hosts Uniref. This database aims to store one representative sequence for each protein without taking into account the species of origin. It clusters all the similar proteins and picks one for every cluster as a representative. There are clusters created at 100%, 90% and 50% identities.

PubMed

PubMed is a bibliographical database that comprises biomedical literature (MEDLINE), life science journals and on-line books. It is a good collection of publications related to biochemistry, cellular biology and medicine. As of 2016 PubMed stores 26 million citations.

For each record it stores:

- Title
- Authors
- Abstract

There is a related database named PubMed Cental (PMC) that only includes citations of Free Access Journals. These citations include the complete text for the papers stored.

PDB, Protein Data Bank

PDB stores 3D structures for proteins and nucleic acids.

Access to the Information in Genbank

Every database provides one or more methods to search and query the data. It is quite common to provide a web interface in which to do text searches with some keyword, author, ID or any other text. Genbank has a powerful query web interface.

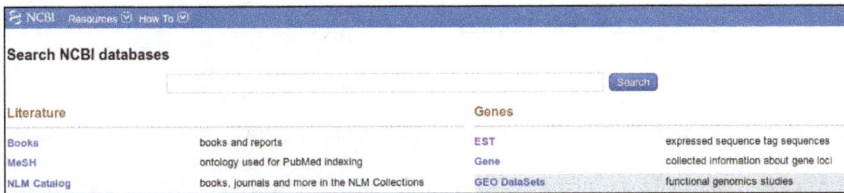

Each database shows the results in one or several formats. For instance, the Genbank sequences can be obtained in several formats.

Genbank format:

```
LOCUS       EC750390                    558 bp    mRNA    linear    EST       03-JUL-2006
DEFINITION  POE00005652 PL(light) Polytomella parva cDNA similar to frataxin protein
                -related, mRNA sequence.
ACCESSION   EC750390
VERSION     EC750390.1  GI:110064507
KEYWORDS    EST.
SOURCE      Polytomella parva
ORGANISM    Polytomella parva
            Eukaryota; Viridiplantae; Chlorophyta; Chlorophyceae;Chlamydomonadales;
                Chlamydomonadaceae; Polytomella.
REFERENCE   1   (bases 1 to 558)
  AUTHORS   Lee,R.W. and Borza,T.
  TITLE     The colorless plastid of the green alga Polytomella parva: a repertoire of
its functions
  JOURNAL   Unpublished (2006)
COMMENT     Contact: TBestDB
            Departement de Biochimie, Universite de Montreal
            Montreal, Canada
            Email: tbestdb-curator@bch.umontreal.ca
            Plate: 4065.
FEATURES             Location/Qualifiers
     source          1..558
                     /organism="Polytomella parva"
```

```
                    /mol_type="mRNA"

                    /db_xref="taxon:51329"

                    /clone_lib="PL(light)"
ORIGIN
        1 gcggccgctt tttttttttt tttttttttt ttttcgtccg ttatttcttt tttaagaatg
       61 cagtcatctg tacatcgtca agtattcgga gtgttatctc gttttgtggg aaacaaagcg
      121 ggtattttta caaagcataa tcatggtgtc tcaaggttgt cttcatgcac ttcgtcatgc
      181 gtaaagatgt atactagcaa caaggccccc gaggatcttc aaacgttcca ccggcaagca
      241 gacgaaactc tagagcaagt cactgaagcc cttgaaaact atgtagatga gcatgaagtg
      301 gaaggcagcg acattgagca tacgcaagga gtgcttacta ttaagcttgg aactcttgga
      361 agttatgtaa ttaataaaca gactcctaat aagcagatat ggttatcctc tcccgtcagt
      421 ggacccttcc gatatgatct taaagaaggt gcctgggttt atgaacgggc tggcgaggct
      481 cggcgcgagc ttatttctca attagaaaca gaaatttcgg atttagttgg tgtcgaatta
      541 aagataagta actgaacg
```

EMBL format:

```
ID    EC750390; SV 1; linear; mRNA; EST; PLN; 558 BP.

XX

AC    EC750390;

XX

DT    04-JUL-2006 (Rel. 88, Created)

DT    04-JUL-2006 (Rel. 88, Last updated, Version 1)

XX

DE    POE00005652 PL(light) Polytomella parva cDNA similar to frataxin

DE    protein-related, mRNA sequence.

XX

KW    EST.

XX

OS    Polytomella parva

OC    Eukaryota; Viridiplantae; Chlorophyta; Chlorophyceae; Chlamydomonadales;

OC    Chlamydomonadaceae; Polytomella.

XX

RN    [1]
```

```
RP    1-558

RA    Lee R.W., Borza T.;

RT    "The colorless plastid of the green alga Polytomella parva: a repertoire of

RT    its functions";

RL    Unpublished.

XX

DR    UNILIB; 42732; 19932.

XX

CC    Contact: TBestDB

CC    Departement de Biochimie, Universite de Montrealhttp://es.wikipedia.org/wiki/Base_
de_datos

CC    Montreal, Canada

CC    Email: tbestdb-curator@bch.umontreal.ca

CC    Plate: 4065.

XX

FH    Key             Location/Qualifiers

FH

FT    source          1..558

FT                    /organism="Polytomella parva"

FT                    /mol_type="mRNA"

FT                    /clone_lib="PL(light)"

FT                    /db_xref="taxon:51329"

FT                    /db_xref="UNILIB:42732"

XX

SQ    Sequence 558 BP; 153 A; 105 C; 127 G; 173 T; 0 other;

      gcggccgctt tttttttttt tttttttttt ttttcgtccg ttatttcttt tttaagaatg        60

      cagtcatctg tacatcgtca agtattcgga gtgttatctc gttttgtggg aaacaaagcg       120

      ggtatttta caaagcataa tcatggtgtc tcaaggttgt cttcatgcac ttcgtcatgc       180

      gtaaagatgt atactagcaa caaggccccc gaggatcttc aaacgttcca ccggcaagca       240

      gacgaaactc tagagcaagt cactgaagcc cttgaaaact atgtagatga gcatgaagtg       300

      gaaggcagcg acattgagca tacgcaagga gtgcttacta ttaagcttgg aactcttgga       360

      agttatgtaa ttaataaaca gactcctaat aagcagatat ggttatcctc tcccgtcagt       420
```

```
    ggacccttcc gatatgatct taaagaaggt gcctgggttt atgaacgggc tggcgaggct        480

    cggcgcgagc ttatttctca attagaaaca gaaatttcgg atttagttgg tgtcgaatta        540

    aagataagta actgaacg                                                      558
```

Main fields in the Genbank format.

Field	Description	Search in Entrez
Locus name	Unique sequence name	[ACCN]
Sequence length	Sequence Length	[SLEN]
Molecule Type	DNA, genomic, mRNA, etc.	[PROP]
Genbank Division	Division for the sequence	[PROP]
Modification Date	Date for the last edit	[MDAT]
Definition	Brief description	[TITL]
Accession	Unique accession ID. It does not changes with modifications	[ACCN]
Version	Version number of the sequence	All fields
Keywords	keywords that describe the sequence	[KYWD]
Source	Common name for the source species	[ORGN]
Organism	Oficial name for the source species	[ORGN]
Reference	Related publications	[TITL] [AUTH] [JOUR]
Features	Regions of interest	[FKEY]
CDS	Coding Sequence	[FKEY]

The Accession is the unique identifier for a sequence record. An accession number applies to the complete record and is usually a combination of a letter(s) and numbers, such as a single letter followed by five digits (e.g., U12345) or two letters followed by six digits (e.g., AF123456).

The records in GenBank can be updated by an author request, accession numbers do not change, even if information in the record is changed. So, a sequence can have several versions in GenBank. Versionis an unique identifier that represents a single, specific sequence in the GenBank database. If there is any change to the sequence data (even a single base), the version number will be increased, e.g., U12345.1 → U12345.2, but the accession portion will remain stable.

Features holds information about genes and gene products, as well as regions of biological significance reported in the sequence. These can include regions of the sequence that code for proteins and RNA molecules.

Features example:

```
FEATURES               Location/Qualifiers
  source               1..12401
                       /organism="Homo sapiens"
                       /mol_type="genomic DNA"
```

```
                    /db_xref="taxon:9606"

                    /chromosome="17"

                    /map="17q24.3"

gene                complement(<1..4774)

                    /gene="SOX9-AS1"

                    /note="SOX9 antisense RNA 1"

                    /db_xref="GeneID:400618"

                    /db_xref="HGNC:HGNC:49321"

ncRNA               complement(<4744..4774)

                    /ncRNA_class="lncRNA"

                    /gene="SOX9-AS1"

                    /product="SOX9 antisense RNA 1, transcript variant 2"

                    /inference="similar to RNA sequence (same

                    species):RefSeq:NR_103737.1"

                    /exception="annotated by transcript or proteomic data"

                    /transcript_id="NR_103737.1"

                    /db_xref="GeneID:400618"

                    /db_xref="HGNC:HGNC:49321"

gene                5001..10401

                    /gene="SOX9"

                    /gene_synonym="CMD1; CMPD1; SRA1; SRXX2; SRXY10"

                    /note="SRY-box 9"

                    /db_xref="GeneID:6662"

                    /db_xref="HGNC:HGNC:11204"

                    /db_xref="MIM:608160"

mRNA                join(5001..5803,6700..6953,7524..10401)

                    /gene="SOX9"

                    /gene_synonym="CMD1; CMPD1; SRA1; SRXX2; SRXY10"

                    /product="SRY-box 9"

                    /transcript_id="NM_000346.3"

                    /db_xref="GeneID:6662"

                    /db_xref="HGNC:HGNC:11204"
```

```
                        /db_xref="MIM:608160"
exon                    5001..5803
                        /gene="SOX9"
                        /gene_synonym="CMD1; CMPD1; SRA1; SRXX2; SRXY10"
                        /inference="alignment:Splign:2.1.0"
                        /number=1
CDS                     join(5373..5803,6700..6953,7524..8368)
                        /gene="SOX9"
                        /gene_synonym="CMD1; CMPD1; SRA1; SRXX2; SRXY10"
                        /note="SRY (sex-determining region Y)-box 9 protein;
                        SRY-related HMG-box, gene 9; SRY (sex determining region
                        Y)-box9"
                        /codon_start=1
                        /product="transcription factor SOX-9"
                        /protein_id="NP_000337.1"
                        /db_xref="CCDS:CCDS11689.1"
                        /db_xref="GeneID:6662"
                        /db_xref="HGNC:HGNC:11204"
                        /db_xref="MIM:608160"
                        /translation="MNLLDPFMKMTDEQEKGLSGAPSPTMSEDSAGSPCPSGSGSDTE
                        NTRPQENTFPKGEPDLKKESEEDKFPVCIREAVSQVLKGYDWTLVPMPVRVNGSSKNK
                        PHVKRPMNAFMVWAQAARRKLADQYPHLHNAELSKTLGKLWRLLNESEKRPFVEEAER
                        LRVQHKKDHPDYKYQPRRRKSVKNGQAEAEEATEQTHISPNAIFKALQADSPHSSSGM
                        SEVHSPGEHSGQSQGPPTPPTTPKTDVQPGKADLKREGRPLPEGGRQPPIDFRDVDIG
                        ELSSDVISNIETFDVNEFDQYLPPNGHPGVPATHGQVTYTGSYGISSTAATPASAGHV
                        WMSKQQAPPPPPQQPPQAPPAPQAPPQPQAAPPQQPAAPPQQPQAHTLTTLSSEPGQS
                        QRTHIKTEQLSPSHYSEQQQHSPQQIAYSPFNLPHYSPSYPPITRSQYDYTDHQNSSS
                        YYSHAAGQGTGLYSTFTYMNPAQRPMYTPIADTSGVPSIPQTHSPQHWEQPVYTQLTR
                        P"
STS                     5374..5670
                        /gene="SOX9"
                        /gene_synonym="CMD1; CMPD1; SRA1; SRXX2; SRXY10"
```

```
                       /standard_name="PMC34415P1"

                       /db_xref="UniSTS:273201"

exon                   6700..6953

                       /gene="SOX9"

                       /gene_synonym="CMD1; CMPD1; SRA1; SRXX2; SRXY10"

                       /inference="alignment:Splign:2.1.0"

                       /number=2

STS                    6765..7647

                       /gene="SOX9"

                       /gene_synonym="CMD1; CMPD1; SRA1; SRXX2; SRXY10"

                       /standard_name="PMC351321P1"

                       /db_xref="UniSTS:273242"

STS                    6771..7605

                       /gene="SOX9"

                       /gene_synonym="CMD1; CMPD1; SRA1; SRXX2; SRXY10"

                       /standard_name="MARC_71570-71571:1249593890:3"

                       /db_xref="UniSTS:528379"

STS                    6876..8104

                       /gene="SOX9"

                       /gene_synonym="CMD1; CMPD1; SRA1; SRXX2; SRXY10"

                       /standard_name="Sox9"

                       /db_xref="UniSTS:502689"

exon                   7524..10401

                       /gene="SOX9"

                       /gene_synonym="CMD1; CMPD1; SRA1; SRXX2; SRXY10"

                       /inference="alignment:Splign:2.1.0"

                       /number=3
```

Sequence Formats

Text Files

There are different formats to store sequences in a text file. Text files should only include Plain text. Graphics or any other binary information are not allowed in text files.

Microsoft Word files are not text files, they are binary files that happen to represent documents. These documents can include text among many other things like images, charts or formats.

Sequences in Plain Files

We could store the sequence in a text file by just writing the sequence. This files would had to include only IUPAC characters.

```
ACAAGATGCCATTGTCCCCCGGCCTCCTGCTGCTGCTGCTCTCCGGGGCCACGGCCACCGCTGCCCTGCCCCTGGAGGGTAC
```

```
GGCCCCACCGGCCGAGACAGCGAGCATATGCAGGAAGCGGCAGGAATAAGGAAAAGCAGCCTCCTGACTTTCCTCGCTTGGT
```

```
AGTGGACCTCCCAGGCCAGTGCCGGGCCCCTCATAGGAGAGGAAGCTCGGGAGGTGGCCAGGCGGCAGGAAGGCGCACCCCC
```

```
ATCCGCGCGCCGGGACAGAATGCCCTGCAGGAACTTCTTCTGGAAGACCTTCTCCTCCTGCAAATAAAA
```

This kind of file is seldom used because it lacks any metadata to identify the sequence.

Fasta Format

The Fasta file includes a name for the sequence and, optionally, some description. The sequence should be preceded by a line that starts with the symbol >. The name will be written after that symbol. Spaces are not allowed in the sequence name. If there is a description it will be found after a space in the same line.

Several sequences can be included in the same file.

It is one of the most common formats.

```
>sequence1_name description
```

```
ACAAGATGCCATTGTCCCCCGGCCTCCTGCTGCTGCTGCTCTCCGGGGCCACGGCCACCGCTGCCCTGCC
```

```
CCTGGAGGGTGGCCCCACCGGCCGAGACAGCGAGCATATGCAGGAAGCGGCAGGAATAAGGAAAAGCAGC
```

```
CTCCTGACTTTCCTCGCTTGGTGGTTTGAGTGGACCTCCCAGGCCAGTGCCGGGCCCCTCATAGGAGAGG
```

```
AAGCTCGGGAGGTGGCCAGGCGGCAGGAAGGCGCACCCCCCCAGCAATCCGCGCGCCGGGACAGAATGCC
```

```
CTGCAGGAACTTCTTCTGGAAGACCTTCTCCTCCTGCAAATAAAACCTCACCCATGAATGCTCACGC
```

```
>sequence2_name description
```

```
ACAAGATGCCATTGTCCCCCGGCCTCCTGCTGCTGCTGCTCTCCGGGGCCACGGCCACCGCTGCCCTGCC
```

```
CCTGGAGGGTGGCCCCACCGGCCGAGACAGCGAGCATATGCAGGAAGCGGCAGGA
```

If we want to include more information we could use the GenBank or EMBL formats. It is also very common in the sequences that come directly from a sequencing machine to include the quality information, for that purpose the most common format is FASTQ.

References

- What-is-bioinformatics: scq.ubc.ca, Retrieved 10 May, 2019
- Sequence-analysis: blueprintgenetics.com, Retrieved 18 March, 2019

- Sequence-alignment, biotech: bioinf.comav.upv.es, Retrieved 7 July, 2019
- Bloom-filters-for-bioinformatics: abhishek-tiwari.com, Retrieved 13 January, 2019
- Structural-annotation, biotech: bioinf.comav.upv.es, Retrieved 3 June, 2019
- Bioinformatics-tutorial: med.cornell.edu, Retrieved 20 February, 2019
- Databases, biotech: comav.upv.es, Retrieved 27 April, 2019

Chapter 2

Computational Methods in Bioinformatics

Computation biology is a domain that is concerned with the development and application of data-analytical and theoretical methods, computational simulation techniques and mathematical modelling, to study biological, ecological, behavioural and social systems. These applications of computational methods have been thoroughly discussed in this chapter.

Application of Machine Learning Methods in Bioinformatics

Machine learning is closely related to computational statistics, which also focuses on prediction-making through the use of computers. It has strong ties to mathematical optimization, which delivers methods, theory and application domains to the field. Machine learning is sometimes conflated with data mining, where the latter subfield focuses more on exploratory data analysis and is known as unsupervised learning. Machine learning can also be unsupervised and be used to learn and establish biology database and help find laws in gene sequences.

Principal Component Analysis

Principal component analysis (PCA) is a statistical procedure that uses an orthogonal transformation to convert a set of observations of possibly correlated variables into a set of values of linearly uncorrelated variables called principal components. Principal component analysis is a statistical method. Through orthogonal transformation, a set of variables that may have correlation is transformed into a set of linearly uncorrelated variables. The transformed set of variables is called principal component. The goal of PCA is to find r, which is used when analyzing complex cubes. When the number of variables is greater than the number of samples, the PCA can minimize the sample dimension to the number of samples without losing the amount of information. It can be seen as a step in the preparation of complex experimental data. The process of PCA is as follows:

- Go to the average, that is, subtract the average for each feature.
- Calculate the covariance matrix.
- Calculate the eigenvalues and eigenvectors of the covariance matrix.
- Sort eigenvalues from largest to smallest.
- Keep the largest number of eigenvectors.
- Transform the data into a new space constructed by a feature vector.

With the help of the machine learning algorithm PCA, we identify five kinds of differentiated alveolar cells and this is shown in figure.

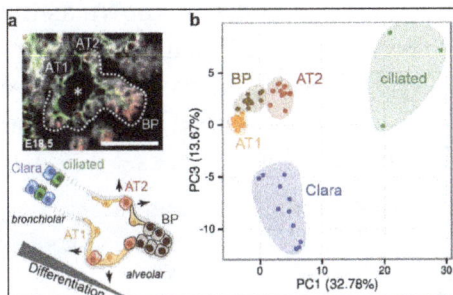

Figure: Distinguish five kinds of cells by the first and third PCA

Back Error Propagation Algotithm

Artificial neural network is a hot research field in recent years. It has always been one of the important research contents in the field of artificial neural network. The basic principle of BP network model processing information is the input signal Xi acts on the output node through an intermediate node (hidden layer point).After nonlinear transformation, it generates an output signal Yk. Each sample trained by the network includes an input vector X and a desired output t, the network output Y, a desired output t by the adjustment of the input node,hidden layer node connection strength W_{ij} value, the hidden layer node, the output node connection strength T_{jk} and threshold, so that the error along the gradient direction is the least. After repeated learning and training to determine, the minimum error corresponding to the network parameters (weights and thresholds) and the training stops. At this point, the trained neural network can process the non-linear transformed information with the smallest output error on the input information. The principle of back error propagation is shown in figure.

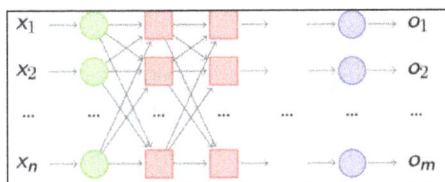

Figure: The principle of BP

In neural networks, neurons receive input signals from other neurons, which are multiplied by weights and added to the total input values received by the neurons, then compared with the current neurons' thresholds and then processed by an activation function, as a result, it generate neuron output. The ideal activation function is a step function, "0" corresponds to neuron depression, and "1" corresponds to neuron excitation. However, the shortcoming of step function is discontinuous, non-conductive and unsmooth, so the sigmoid function is often used as an activation function instead of a step function.

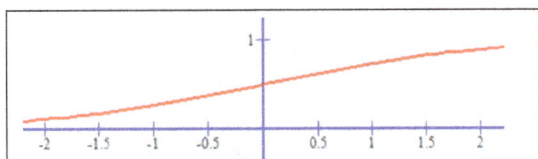

Figure: Sigmoid function

Protein secondary structure prediction is based on the known primary structure, the prediction methods and techniques to achieve the classification of secondary structure prediction. For BP neural network, the input is the primary sequence of known as the protein, the output is the secondary structure type. In this paper, a 3-layer BP neural network is used as a classifier to predict the protein secondary structure. Design, the BP network input layer is designed to slide along the amino acid sequence of the window, the window position is symmetrical. The disadvantage of this algorithm is that it can be overfit and needs lots of tricks.

Neural Networks

Neural neworks are typically organized in layers. Layers are made up of a number of interconnected 'nodes' which contain an 'activation function'. Patterns are presented to the network via the 'input layer', which communicates to one or more 'hidden layers' where the actual processing is done via a system of weighted 'connections'. The hidden layers then link to an 'output layer' where the answer is output as shown in the graphic below.

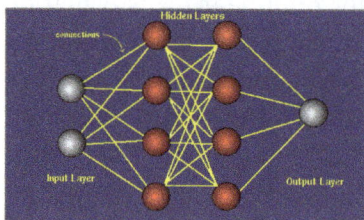

Most ANNs contain some form of 'learning rule' which modifies the weights of the connections according to the input patterns that it is presented with. In a sense, ANNs learn by example as do their biological counterparts; a child learns to recognize dogs from examples of dogs.

Although there are many different kinds of learning rules used by neural networks, this demonstration is concerned only with one; the delta rule. The delta rule is often utilized by the most common class of ANNs called 'backpropagational neural networks' (BPNNs). Backpropagation is an abbreviation for the backwards propagation of error.

With the delta rule, as with other types of backpropagation, 'learning' is a supervised process that occurs with each cycle or 'epoch' (i.e. each time the network is presented with a new input pattern) through a forward activation flow of outputs, and the backwards error propagation of weight adjustments. More simply, when a neural network is initially presented with a pattern it makes a random 'guess' as to what it might be. It then sees how far its answer was from the actual one and makes an appropriate adjustment to its connection weights. More graphically, the process looks something like this:

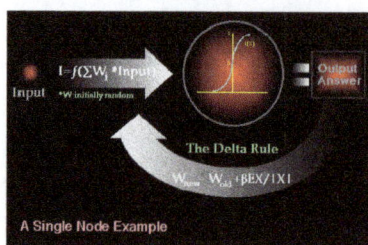

That within each hidden layer node is a sigmoidal activation function which polarizes network activity and helps it to stablize.

Backpropagation performs a gradient descent within the solution's vector space towards a 'global minimum' along the steepest vector of the error surface. The global minimum is that theoretical solution with the lowest possible error. The error surface itself is a hyperparaboloid but is seldom 'smooth' as is depicted in the graphic below. Indeed, in most problems, the solution space is quite irregular with numerous 'pits' and 'hills' which may cause the network to settle down in a local.

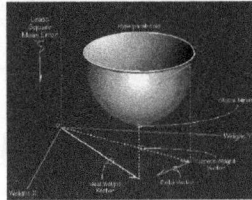

Minum' which is not the best overall solution

Since the nature of the error space can not be known a prioi, neural network analysis often requires a large number of individual runs to determine the best solution. Most learning rules have built-in mathematical terms to assist in this process which control the 'speed' (Beta-coefficient) and the 'momentum' of the learning. The speed of learning is actually the rate of convergence between the current solution and the global minimum. Momentum helps the network to overcome obstacles (local minima) in the error surface and settle down at or near the global miniumum.

Once a neural network is 'trained' to a satisfactory level it may be used as an analytical tool on other data. To do this, the user no longer specifies any training runs and instead allows the network to work in forward propagation mode only. New inputs are presented to the input pattern where they filter into and are processed by the middle layers as though training were taking place, however, at this point the output is retained and no backpropagation occurs. The output of a forward propagation run is the predicted model for the data which can then be used for further analysis and interpretation.

It is also possible to over-train a neural network, which means that the network has been trained exactly to respond to only one type of input; which is much like rote memorization. If this should happen then learning can no longer occur and the network is refered to as having been "grandmothered" in neural network jargon. In real-world applications this situation is not very useful since one would need a separate grandmothered network for each new kind of input.

How do Neural Networks Differ from Conventional Computing?

To better understand artificial neural computing it is important to know first how a conventional 'serial' computer and it's software process information. A serial computer has a central processor that can address an array of memory locations where data and instructions are stored. Computations are made by the processor reading an instruction as well as any data the instruction requires from memory addresses, the instruction is then executed and the results are saved in a specified memory location as required. In a serial system (and a standard parallel one as well) the computational steps are deterministic, sequential and logical, and the state of a given variable can be tracked from one operation to another.

In comparison, ANNs are not sequential or necessarily deterministic. There are no complex central processors, rather there are many simple ones which generally do nothing more than take the weighted sum of their inputs from other processors. ANNs do not execute programed instructions; they respond in parallel (either simulated or actual) to the pattern of inputs presented to it. There are also no separate memory addresses for storing data. Instead, information is contained in the overall activation 'state' of the network. 'Knowledge' is thus represented by the network itself, which is quite literally more than the sum of its individual components.

What Applications Should Neural Networks be used for?

Neural networks are universal approximators, and they work best if the system you are using them to model has a high tolerance to error. One would therefore not be advised to use a neural network to balance one's cheque book. However they work very well for:

- Capturing associations or discovering regularities within a set of patterns;

- Where the volume, number of variables or diversity of the data is very great;

- The relationships between variables are vaguely understood;

- The relationships are difficult to describe adequately with conventional approaches.

What are their Limitations?

There are some specific issues potential users should be aware of:

- Backpropagational neural networks (and many other types of networks) are in a sense the ultimate 'black boxes'. Apart from defining the general archetecture of a network and perhaps initially seeding it with a random numbers, the user has no other role than to feed it input and watch it train and await the output. In fact, it has been said that with backpropagation, "you almost don't know what you're doing". Some software freely available software packages (NevProp, bp, Mactivation) do allow the user to sample the networks 'progress' at regular time intervals, but the learning itself progresses on its own. The final product of this activity is a trained network that provides no equations or coefficients defining a relationship (as in regression) beyond it's own internal mathematics. The network 'IS' the final equation of the relationship.

- Backpropagational networks also tend to be slower to train than other types of networks and sometimes require thousands of epochs. If run on a truly parallel computer system this issue is not really a problem, but if the BPNN is being simulated on a standard serial machine (i.e. a single SPARC, Mac or PC) training can take some time. This is because the machines CPU must compute the function of each node and connection separately, which can be problematic in very large networks with a large amount of data. However, the speed of most current machines is such that this is typically not much of an issue.

What are their Advantages over Conventional Techniques?

Depending on the nature of the application and the strength of the internal data patterns you can generally expect a network to train quite well. This applies to problems where the relationships may be quite dynamic or non-linear. ANNs provide an analytical alternative to conventional techniques which are often limited by strict assumptions of normality, linearity, variable independence etc. Because an ANN can capture many kinds of relationships it allows the user to quickly and relatively easily model phenomena which otherwise may have been very difficult or imposible to explain otherwise.

Deep Neural Networks

The basic structure of DNNs consists of an input layer, multiple hidden layers, and an output layer. Once input data are given to the DNNs, output values are computed sequentially along the layers of the network. At each layer, the input vector comprising the output values of each unit in the layer below is multiplied by the weight vector for each unit in the current layer to produce the weighted sum. Then, a nonlinear function, such as a sigmoid, hyperbolic tangent, or rectified linear unit (ReLU), is applied to the weighted sum to compute the output values of the layer. The computation in each layer transforms the representations in the layer below into slightly more abstract representations. Based on the types of layers used in DNNs and the corresponding learning method, DNNs can be classified as MLP, SAE, or DBN.

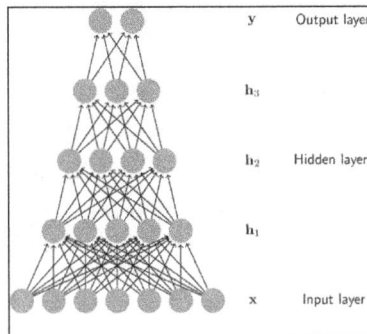

MLP has a similar structure to the usual neural networks but includes more stacked layers. It is trained in a purely supervised manner that uses only labeled data. Since the training method is a process of optimization in high-dimensional parameter space, MLP is typically used when a large number of labeled data are available.

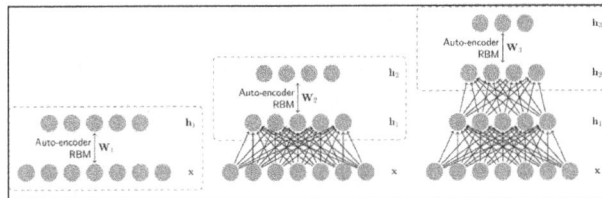

SAE and DBN use AEs and RBMs as building blocks of the architectures, respectively. The main difference between these and MLP is that training is executed in two phases: unsupervised pre-training and supervised fine-tuning. First, in unsupervised pre-training, the layers are stacked sequentially and trained in a layer-wise manner as an AE or RBM using unlabeled data. Afterwards, in supervised fine-tuning, an output classifier layer is stacked, and the whole neural network is optimized by retraining with labeled data. Since both SAE and DBN exploit unlabeled data and can help avoid overfitting, researchers are able to obtain fairly regularized results, even when labeled data are insufficient as is common in the real world.

DNNs are renowned for their suitability in analyzing high-dimensional data. Given that bioinformatics data are typically complex and high-dimensional, DNNs have great promise for bioinformatics research. We believe DNNs, as hierarchical representation learning methods, can discover previously unknown highly abstract patterns and correlations to provide insight to better understand the nature of the data. However, it has occurred to us that the capabilities of DNNs have not yet fully been exploited. Although the key characteristic of DNNs is that hierarchical features are learned solely from data, human designed features have often been given as inputs instead of raw data forms. We expect that the future progress of DNNs in bioinformatics will come from investigations into proper ways to encode raw data and learn suitable features from them.

Convolutional Neural Networks

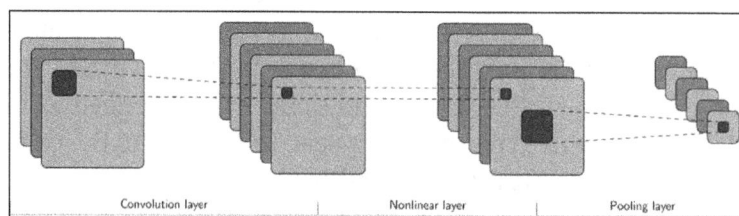

| Convolution layer | Nonlinear layer | Pooling layer |

CNNs are designed to process multiple data types, especially two-dimensional images, and are directly inspired by the visual cortex of the brain. In the visual cortex, there is a hierarchy of two basic cell types:simple cells and complex cells. Simple cells react to primitive patterns in sub-regions of visual stimuli, and complex cells synthesize the information from simple cells to identify more intricate forms. Since the visual cortex is such a powerful and natural visual processing system, CNNs are applied to imitate three key ideas: local connectivity, invariance to location, and invariance to local transition.

The basic structure of CNNs consists of convolution layers, nonlinear layers, and pooling layers. To use highly correlated sub-regions of data, groups of local weighted sums, called feature maps, are obtained at each convolution layer by computing convolutions between local patches and weight vectors called filters. Furthermore, since identical patterns can appear.

Regardless of the location in the data, filters are applied repeatedly across the entire dataset, which also improves training efficiency by reducing the number of parameters to learn. Then nonlinear layers increase the nonlinear properties of feature maps. At each pooling layer, maximum or average subsampling of non-overlapping regions in feature maps is performed. This non-overlapping subsampling enables CNNs to handle somewhat different but semantically similar features and thus aggregate local features to identify more complex features.

Currently, CNNs are one of the most successful deep learning architectures owing to their outstanding capacity to analyze spatial information. Thanks to their developments in the field of object recognition, we believe the primary research achievements in bioinformatics will come from the biomedical imaging domain. Despite the different data characteristics between normal and biomedical imaging, CNN will nonetheless offer straightforward applications compared to other domains. Indeed, CNNs also have great potential in omics and biomedical signal processing. The three keys ideas of CNNs can be applied not only in a one-dimensional grid to discover meaningful recurring patterns with small variance, such as genomic sequence motifs, but also in two-dimensional grids, such as interactions within omics data and in timefrequency matrices of biomedical signals. Thus, we believe the popularity and promise of CNNs in bioinformatics applications will continue in the years ahead.

Recurrent Neural Networks

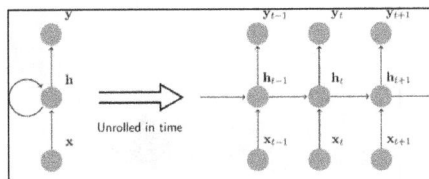

RNNs, which are designed to utilize sequential information, have a basic structure with a cyclic connection. Since input data are processed sequentially, recurrent computation is performed in the hidden units where cyclic connection exists. Therefore, past information is implicitly stored in the hidden units called state vectors, and output for the current input is computed considering all previous inputs using these state vectors. Since there are many cases where both past and future inputs affect output for the current input (e.g., in speech recognition), bidirectional recurrent neural networks (BRNNs) have also been designed and used widely.

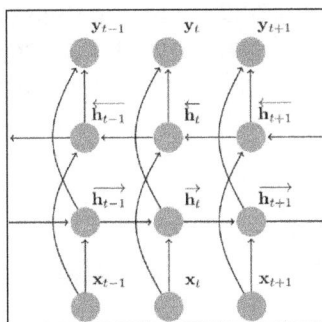

Although RNNs do not seem to be deep as DNNs or CNNs in terms of the number of layers, they can be regarded as an even deeper structure if unrolled in time. Therefore, for a long time, researchers struggled against vanishing gradient problems while training RNNs, and learning long-term dependency among data was difficult. Fortunately, substituting the simple perceptron hidden units with more complex units such as LSTM or GRU , which function as memory cells, significantly helps to prevent the problem. More recently, RNNs have been used successfully in many areas including natural language processing and language translation.

Even though RNNs have been explored less than DNNs and CNNs, they still provide very powerful analysis methods for sequential information. Since omics data and biomedical signals are typically sequential and often considered languages of nature, the capabilities of RNNs for mapping a variable-length input sequence to another sequence or fixed-size prediction are promising for bioinformatics research. With regard to biomedical imaging, RNNs are currently not the first choice of many researchers. Nevertheless, we believe that dissemination of dynamic CT and MRI would lead to the incorporation of RNNs and CNNs and elevate their importance in the long term. Furthermore, we expect that their successes in natural language processing will lead RNNs to be applied in biomedical text analysis and that employing an attention mechanism will improve performance and extract more relevant information from bioinformatics data.

Emergent Architectures

Emergent architectures refer to deep learning architectures besides DNNs, CNNs, and RNNs. In this review, we introduce three emergent architectures (i.e., DST-NNs, MD-RNNs, and CAEs) and their applications in bioinformatics.

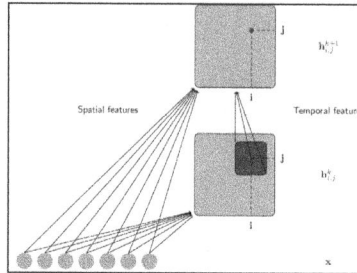

DST-NNs are designed to learn multi-dimensional output targets through progressive refinement. The basic structure of DST-NNs consists of multi-dimensional hidden layers. The key aspect of the structure, progressive refinement, considers local correlations and is performed via input feature compositions in each layer: spatial features and temporal features. Spatial features refer to the original inputs for the whole DST-NN and are used identically in every layer. However, temporal features are gradually altered so as to progress to the upper layers. Except for the first layer, to compute each hidden unit in the current layer, only the adjacent hidden units of the same coordinate in the layer below are used so that local correlations are reflected progressively.

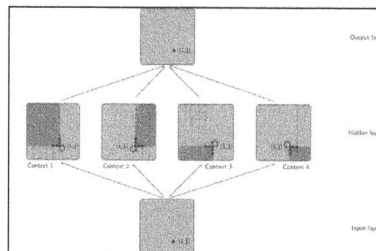

MD-RNNs are designed to apply the capabilities of RNNs to non-sequential multidimensional data by treating them as groups of sequential data. For instance, two-dimensional data are treated as groups of horizontal and vertical sequence data. Similar to BRNNs which use contexts in both directions in one-dimensional data, MD-RNNs use contexts in all possible directions in the multi-dimensional data. In the example of a two-dimensional dataset, four contexts that vary with the order of data processing are reflected in the computation of four hidden units for each position in the hidden layer. The hidden units are connected to a single output layer, and the final results are computed with consideration of all possible contexts.

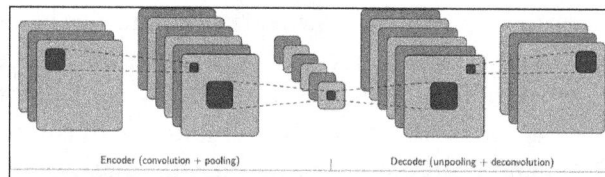

CAEs are designed to utilize the advantages of both AE and CNNs so that it can learn good hierarchical representations of data reflecting spatial information and be well regularized by unsupervised training. In training of AEs, reconstruction error is minimized using an encoder and decoder, which extract feature vectors from input data and recreate the data from the feature vectors, respectively. In CNNs, convolution and pooling layers can be regarded as a type of encoder. Therefore, the CNN encoder and decoder consisting of deconvolution and unpooling layers are integrated to form a CAE and are trained in the same manner as in AE.

Deep learning is a rapidly growing research area, and a plethora of new deep learning architecture is being proposed but awaits wide applications in bioinformatics. Newly proposed architectures have different advantages from existing architectures, so we expect them to produce promising results in various research areas. For example, the progressive refinement of DST-NNs fits the dynamic folding process of proteins and can be effectively utilized in protein structure prediction; the capabilities of MD-RNNs are suitable for segmentation of biomedical images since segmentation requires interpretation of local and global contexts; the unsupervised representation learning with consideration of spatial information in CAEs can provide great advantages in discovering recurring patterns in limited and imbalanced bioinformatics data.

Artificial Neural Networks

ANN stands for Artificial Neural Networks. Basically, it's a computational model. That is based on structures and functions of biological neural networks. Although, the structure of the ANN affected by a flow of information. Hence, neural network changes were based on input and output.

Basically, we can consider ANN as nonlinear statistical data. That means complex relationship defines between input and output. As a result, we found different patterns. Also, we call the ANN as a neural network.

Structure of Artificial Neural Network

Generally, the working of a human brain by making the right connections is the idea behind ANNs. That was limited to use of silicon and wires as living neurons and dendrites.

Here, neurons, part of human brain. That was composed of 86 billion nerve cells. Also, connected to other thousands of cells by Axons. Although, there are various inputs from sensory organs. That was accepted by dendrites. As a result, it creates electric impulses. That is used to travel through the Artificial neural network. Thus, to handle the different issues, neuron send a message to another neuron.

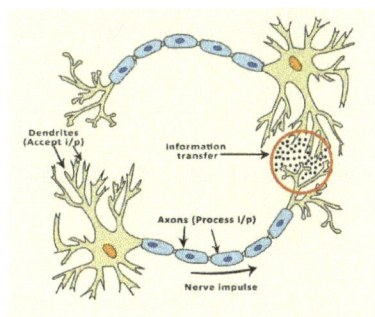

Basic Structure of Artificial Neural Network

As a result, we can say that ANNs are composed of multiple nodes. That imitate biological neurons of the human brain. Although, we connect these neurons by links. Also, they interact with each other. Although, nodes are used to take input data. Further, perform simple operations on the data. As a result, these operations are passed to other neurons. Also, output at each node is called its activation or node value.

As each link is associated with weight. Also, they are capable of learning. That takes place by altering weight values. Hence, the following illustration shows a simple ANN.

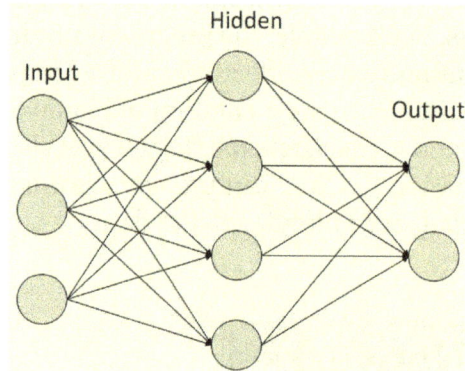

Artificial Neural Network Structure

Types of Artificial Neural Networks

Generally, there are two types of ANN. Such as FeedForward and Feedback.

a. FeedForward ANN

In this network flow of information is unidirectional. A unit used to send information to another unit that does not receive any information. Also, no feedback loops are present in this. Although, used in recognition of a pattern. As they contain fixed inputs and outputs.

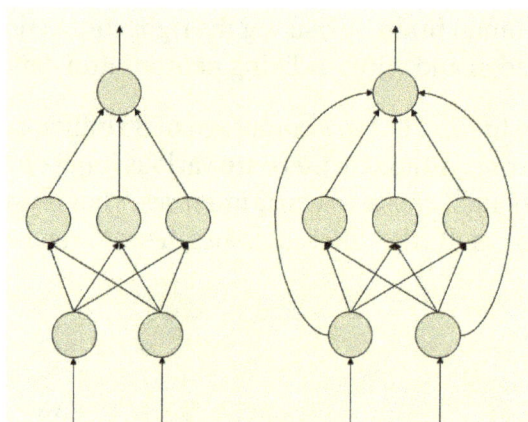

Types of Artificial Neural Networks – FeedForward ANN

b. FeedBack ANN

In this particular Artificial Neural Network, it allows feedback loops. Also, used in content addressable memories.

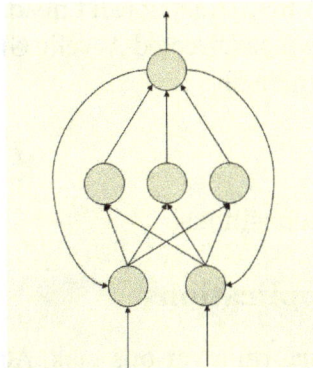

Types of Artificial Neural Networks – FeedBack ANN

How Does Artificial Neural Networks Works?

In this topology diagrams, you will learn everything in a detailed manner.

In this, each arrow represents a connection between two neurons. Also, they used to indicate the pathway for the flow of information. As it was noticed that each connection has a weight, an integer number. That used to controls the signal between the two neurons.

If the output is good that was generated by the network then we don't require to adjust the weights. Although, if poor output generated. Then surely system will alter the weight to improve results.

Machine Learning in ANNs

As there are too many Machine learning strategies are present, let's see them one by one:

a. Supervised Learning: Generally, in this learning a teacher is present to teach. That teacher must be aware of ANN. For example: The teacher feeds only example data. That teacher already knows the answers.

b. Unsupervised Learning: If there is present no data set. Then we need this learning technique.

c. Reinforcement Learning: As this Machine learning technique is based on the observation. Although, if it's negative the networks need to adjust its weights. That is able to make a different required decision the next time.

Back Propagation Algorithm

Generally, we use to call it as training and learning algorithm. As these networks are ideal for simple Pattern Recognition and Mapping Task.

Bayesian Networks

Basically, we use to call it as graphical structures. Generally, we use this network to represent probabilistic representation. This represents among a set of random variables. Also, we used to call this network as Belief networks or Bayes Nets.

In these networks, each node represents a random variable with specific propositions.

In this only constraint arcs present in BN. Thus, doesn't need to return node by following directed arcs. Hence, we can say BNs are known as Directed Acyclic Graphs (DAGs). Hence, we use BNs to handle multivalued variables simultaneously.

Range of Prepositions

Probability assigned to each of the prepositions.

Artificial Neural Networks Applications

Artificial Neural Network used to perform a various task. Also, this task performs that are busy with humans but difficult for a machine.

a. Aerospace: Generally, we use ANN a for Autopilot aircrafts. They used for aircraft fault detection.

b. Military: In various ways, we use ANN an in the military. Such as Weapon orientation and steering, target tracking.

c. Electronics: Basically, we use an Artificial neural network in electronics in many ways. That are code sequence prediction, IC chip layout, and chip failure analysis.

d. Medical: As medical has too many machines. That use in various ways. Such as cancer cell analysis, EEG and ECG analysis.

e. Speech: We use ANN in speech recognition and speech classification.

f. Telecommunications: Generally, it has different applications. Thus, we use an Artificial neural network in many ways. Such as image and data compression, automated information services.

g. Transportation: Generally, we use an Artificial neural network in transportation in many ways. That are truck Brake system diagnosis and vehicle scheduling, routing systems.

h. Software: It also uses an ANN in pattern Recognition. Such as in facial recognition, optical character recognition, etc.

i. Time Series Prediction: We use an Artificial neural network to predict time. Also, we use ANNs to make predictions on stocks and natural calamities.

Bioinformatics Algorithms

Genetic Algorithm

When it comes to bioinformatics algorithms, Genetic algorithms top the list of most used and talked about algorithms in bioinformatics. Understanding the Genetic algorithm is important not only because it helps you to reduce the computational time taken to get a result but also because it is inspired by how nature works.

Genetic Algorithm was developed by John Holland. It uses the concepts of Natural Selection and Genetic Inheritance and tries to mimic the biological evolution. It falls under the category of algorithms known as Evolutionary Algorithms. It can be used to find a solution to the hard problems where we don't know much about the search space.

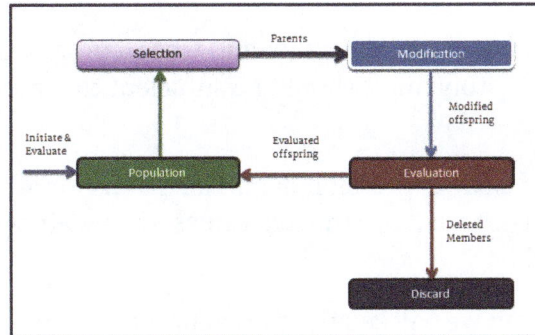

Let us understand how a genetic algorithm works. For this, let us consider a cancer-associated gene expression matrix. This matrix contains all the known genes found in human being and their level of expression.

- For a given problem, the genetic algorithm works by maintaining a set of candidate solutions and then applies three operators over them – Selection, Recombination, and Mutation, which are collectively known as the stochastic operators.

- Selection: In nature, if an organism is adapted to the environment, its population will grow relative to its quality of adaptation. This is referred to as selection. It means if a solution meets the conditional constraints, it is replicated at a rate which is proportional to the relative quality.

- Recombination: In nature, two similar chromosomes of the surviving individual exchange genes during sexual reproduction in a process known as Crossing Over. In GA we decompose two distinct solutions and randomly mix their parts to form novel solutions.

- Mutation: Random changes in an existing chromosome may lead to some fitter individual. This concept is utilized to randomly perturbs a candidate solution:

  1. produce an initial population of individuals.

  2. evaluate the fitness of all individuals.

  3. while termination condition not met do.

  4. select fitter individuals for reproduction.

  5. recombine between individuals.

  6. mutate individuals.

  7. evaluate the fitness of the modified individuals.

8. generate a new population.

9. End while.

The Program

We are going to implement the Genetic Algorithm and write a program in Perl for it. Although not purely applicable to a real-life problem, it should be sufficient to familiarize you with the Genetic Algorithm.

Suppose that you had a set of Gene expression data. The data is for all 25000 genes in the human genome and you want to find out what are the five values among all 25000 values whose sum can give you the highest number.

For the purpose of this program we will require four subroutines:

- Generate: It will generate chromosomes containing 5 values(specified in variable $GeneNumberConstraint) selected at random at positions.

- Mutate: It mutates a chromosome at random position with a random value less than specified in $HighestMutationValue.

- Survival Check: It checks if the newly formed chromosome is viable. i.e. It has a value that is up to a minimum specification. (Checking for fitness).

- Recombine: It will form new combinations from the existing chromosome by crossing them over with each other.

The Code

So here is the final code implementing Genetic Algorithm in Perl:

```
$CurrentHighest=0;

@GeneExpressionData = (1,3,8,5,2,4,46,6,78,7,9,9

,0,1,1,1,5,59,9,97,7,6,5,45

,4,3,23,2,22,2,2,4,5,5,6,54);

@SolutionSpace = ();

$HighestMutationValue = 110;

$GeneNumberConstraint = 5;

$InitialThreshold  = 10;

$genes = scalar @GeneExpressionData;

@chromosome = ();

 $sum = 0;

$steps= 10;

print "The Total Genes are: $genes\n";
```

```perl
generate();

$steps = 10;

for($p=0;$p&lt;=$steps;$p++){

 generate();

 SurvivalCheck();

 mutate();

 SurvivalCheck();

 recombine();

 SurvivalCheck();

}

print "\n\n Genetic Algorithm Result

\n\n\n\t\tHighest Detected: $CurrentHighest in $steps Steps\n\n";

sub mutate{

 $randpos = int(rand($gene));

 $n = int(rand($HighestMutationValue));

 $chromosome[$randpos] = $n;

 print "\n Mutation Took Place in Chromosome @chromosome ";

}

sub recombine

{

   print "\nRecombining\n\n";

   @chromosome1 = $SolutionSpace[int rand($p)];

   @chromosome2 =  $SolutionSpace[int rand($p)];

  print "Random Sequence Chromosome from Solution Space: @chromosome1 and @chromosome2";

   for($i=0; $i&lt;=$GeneNumberConstraint/2; $i++){ my $random_number = int(rand(3)) +
1; $pos1 = int(rand($GeneNumberConstraint)); $pos1 = int(rand($GeneNumberConstraint));
$swap = $chromosome1[$pos1]; $chromosome1[$pos1] = $chromosome2[$pos2]; $chromosome2[$-
pos2] = $swap; } print "The Recombination led to @chromosome"; @chromosome = (); @chro-
mosome = @chromosome1; } sub SurvivalCheck{ $sum = 0; foreach $val (@chromosome){ $sum
+= $val; } if($sum&gt;$CurrentHighest){

   $CurrentHighest = $sum;

   push @SolutionSpace, @chromosome;

   print "\nIndividual is alive! \nCurrent Highest Expression: $CurrentHighest";
return 1;

 }

 else{

  print "\nSpecies Didn't Survive! \n";
```

```perl
    return 0;
  }
}

sub generate{
 @chromosome = ();
 for($i=1;$i&lt;=$GeneNumberConstraint;$i++){
  $n = int(rand($genes));
  push @chromosome, $GeneExpressionData[$n];
  $sum += $GeneExpressionData[$n];
 }
 print "\n\n\nGenerated Chromosome: @chromosome \n";
}
```

Smith-Waterman Algorithm

The Smith–Waterman algorithm is a well-known algorithm for performing local sequence alignment; that is, for determining similar regions between two nucleotide or protein sequences. Instead of looking at the total sequence, the Smith–Waterman algorithm compares segments of all possible lengths and optimizes the similarity measure. This algorithm is a variation of Needleman-Wunsch Algorithm developed by Temple F. Smith and Michael S. Waterman in 1981, it is also a dynamic programming algorithm to find the optimal local alignment with respect to the scoring system being used. The major difference from Needleman-Wunsch algorithm includes:

- In order to highlight the best local alignments; negative scoring matrix cells are set to zero.

- Traceback procedure starts at the highest scoring matrix cell and proceeds until a cell with score zero is encountered.

This algorithm was designed to sensitively detect similarities in highly diverged sequences. For the purpose of explanation, first summarizing the algorithm in to simple steps and then you will be move forwards with examples and explanations.

Algorithm is similar to global alignment with modified boundary conditions and recurrence rules,

1. The top row and left column are now filled with 0

2. If the (sub-)alignment score becomes negative, restart the search:

$$F(i,j) = \max \begin{cases} 0, \\ F(i-1,j-1) + s(x_i, y_j), \\ F(i-1,j) - d, \\ F(i,j-1) - d. \end{cases}$$

1. Traceback is from the maximum of F (i, j) in the whole matrix to the first 0.

2. Example: the optimal local alignment between HEAGAWGHEE and PAWHEAE is AW-GHE::AW-HE.

3. Issue: In gapped alignments, the expected score for a random match may be positive

Example of Smith–Waterman Algorithm:

1. Start with a N × N integer matrix where N is sequence length (both s and t). Compute M[i][j] based on Score Matrix and optimum score compute so far (DP).

	0	C	G	G	G	T	A	T	C	C	A	A
0												
C												
C												
C												
T												
A												
G												
G												
T												
C												
C												
C												
C												

2. Understanding the matrix

- Alignment
  ```
  t:- - - - - - - -
  s:ccctaggt
  ```

	0	C	G	G	G	T	A	T	C	C	A	A
0	0											
C	0											
C	0											
C	0											
T	0											
A	0											
G	0											
G	0											
T	0											
C	0											
C	0											
C	0											
C	0											

- Alignment
  ```
  t:cgggtat...
  s:- - - - - - -...
  ```

	0	C	G	G	G	T	A	T	C	C	A	A
0	0	0	0	0	0	0	0	0	0	0	0	0
C	0											
C	0											
C	0											
T	0											
A	0											
G	0											
G	0											
T	0											
C	0											
C	0											
C	0											
C	0											

3. Computing cell scores: Finding m[i][j]: There are three ways to finish the alignment of so..i and to j:

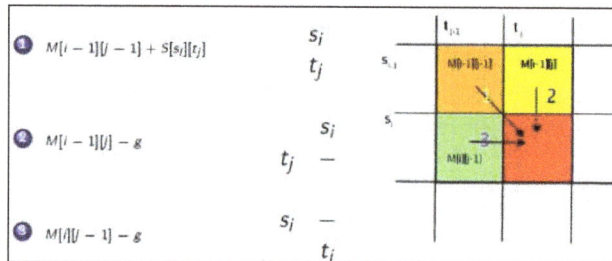

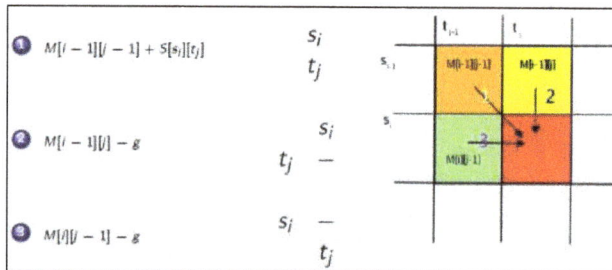

4. Scoring process: Element Computation M[i][j]:

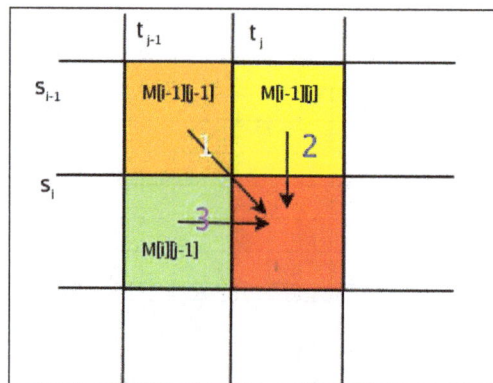

$$M[i][0]=0 \quad M[0][j]=0$$

$$M[0][j]=\max \begin{cases} 0 \\ M[i-1][j-1]+S[s_i][t_j] & \text{if } s_i\ t_j \\ M[i-1][j]-d & \text{if } s_i, \\ M[i][j-1]-d & \text{if } t_j, \end{cases}$$

5. Backtracking Process: For finding the BEST local alignment, find the highest score and then traceback to first 0.

Scoring used in this example:

- Diag - the letters from two sequences are aligned.

- Left - a gap is introduced in the left sequence.

- Up - a gap is introduced in the top sequence.

$$\text{Score}_{opt} = \max_{i,j=1}^{N} M[i][j]$$

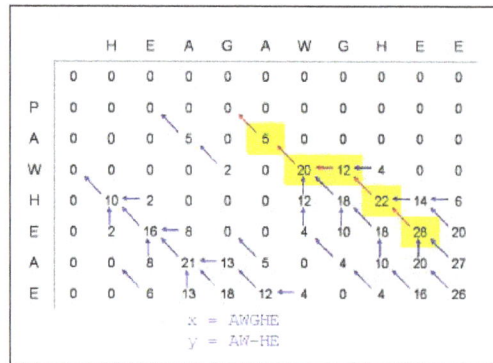

	H	E	A	G	A	W	G	H	E	E
	0	0	0	0	0	0	0	0	0	0
P	0	0	0	0	0	0	0	0	0	0
A	0	0	0	5	0	5	0	0	0	0
W	0	0	0	0	2	0	20	12	4	0	0
H	0	10	2	0	0	0	12	18	22	14	6
E	0	2	16	8	0	0	4	10	18	28	20
A	0	0	8	21	13	5	0	4	10	20	27
E	0	0	6	13	18	12	4	0	4	16	26

x = AWGHE
y = AW-HE

Probabilistic Model for Sequence Analysis

A probabilistic model is a model that produces different outcomes with different probabilities – it can simulate a whole class of objects, assigning each an associated probability. In bioinformatics, the objects usually are DNA or protein sequences and a model might describe a family of related sequences.

A DNA sequence is a succession of the letters A, C, T and G. The sequences are any combination of these letters. A physical or mathematical model of a system produces a sequence of symbols according to a certain probability associated with them. This is known as a stochastic process, that is, it is a mathematical model for a biological system which is governed by a set of probability measure. The occurrence of the letters can lead us to the further study of genetic disorder. The stochastic process is also known mathematically as Discrete Markov Process. There are different ways to use probabilities for depicting the DNA sequences. One scheme is where each occurrence of a letter is independent of the occurrence of any other letter. That is, the probability of occurrence of a letter does not depend on the occurrence of the previous one. In yet another scheme, the occurrence of a letter depends on the occurrence of the previous one. The earlier one is termed in as unsuccessive probability and the later one is termed as successive probability. From this study we can further show the relationship between sequence and genetic variations. It can also lead to a more powerful test for identifying particular classes of genes or proteins which has been illustrated by an example.

DNA Sequence

DNA sequences is a succession of letters representing the primary structure of a real or hypothetical DNA molecule or strand, with the capacity to carry information as described by the central dogma of molecular biology. There are 4 nucleotide bases (A – Adenine, C – Cytosine, G – Guanine, T – Thymine). DNA sequencing is the process of determining the exact order of the bases A, T, C and G in a piece of DNA. In essence, the DNA is used as a template to generate a set of fragments that differ in length from each other by a single base. The fragments are then separated by size, and the bases at the end are identified, recreating the original sequence of the DNA. The most commonly used method of sequencing DNA the dideoxy or chain termination method was developed by Fred Sanger in 1977 (for which he won his second Nobel Prize). The key to the method is the use of modified bases called dideoxy bases; when a piece of DNA is being replicated and a dideoxy base is incorporated into the new chain, it stops the replication reaction.

Most DNA sequencing is carried out using the chain termination method. This involves the synthesis of new DNA strands on a single standard template and the random incorporation of chain-terminating nucleotide analogues. The chain termination method produces a set of DNA molecules differing in length by one nucleotide. The last base in each molecule can be identified by way of a unique label. Separation of these DNA molecules according to size places them in correct order to read off the sequence.

Different Probabilistic Approaches for Sequence Representation

A DNA sequence is essentially represented as a string of four characters A, C, T, G and looks something like ACCTGACCTTACG. These strings can also be represented in terms of some probability measures and using these measures it can depicted graphically as well. This graphical representation matches the Markov Hidden Model. Some of these schemes are presented in this paper. A physical or mathematical model of a system produces a sequence of symbols according to a certain probability associated with them. This is known as a stochastic process. There are different ways to use probabilities for depicting the DNA sequences.

Unsuccessive Probability

In this representation scheme the probability of the next occurrence of a letter does not depend on the previous letter. There can be two different representation schemes in this: one which uses equal probability for each letter and another which assumes a fixed probability for each letter.

- Equal Probability - Suppose we have a 4 letter code consisting of the 4 letters A, C, T, G which can be chosen each with equal probability 0.25, successive choices being independent. This would lead to a sequence of which the following is a typical example:

  ACCATGGACTTAGCTACTGG

- Unequal Probability - For a DNA sequence series of length 20, where each letter has a probability of 0.3, 0.2, 0.3, 0.2 respectively, with successive choices are independent. A typical message from this source is:

  ACTTGAAATTCGGACCTGAT

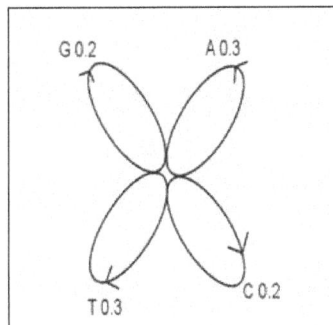

Figure: Graphical Representation of Unequal Probability Sequence

Successive Probability

This section presents two schemes for representing a DNA sequence: one where successive letter depends on the preceding one and in another scheme the letters are used to form words and a probability is associated with each word.

- Letter Probability - A more complicated structure is obtained if successive symbols are not chosen independently but their probabilities depend on preceding letters. In the simplest case of this type a choice depends only on the preceding letter and not on ones before that. The statistical structure can then be described by a set of transition probabilities $pi(j)$, the probability that a letter i is followed by letter j. The indices i and j range over all the possible symbols. A second equivalent way of specifying the structure is to give the "digram" probabilities $p(i, j)$, i.e., the relative frequency of the digram ij. The letter frequencies $p(i)$, (the probability of letter i), the transition probabilities $pi(j)$ and the digram probabilities $p(i, j)$ are related by the following formulas:

$$p(i) = \sum_j p(i,j) = \sum_j p(j,i) = \sum_j p(j) p_j$$

$$p(i,j) = p(i) p_i(j)$$

$$\sum_j p_i(j) = \sum_i p(i) = \sum_{i,j} p(i,j) = 1$$

In the example we have used the letter probability for representing the sequence.

- Word Probability - A process can also be defined which produces a text consisting of a sequence of words. Since a DNA sequence consists of typically 4 letters A, C, T, G any word could be made up of these four letters. For example the typical DNA sequence may consist of any combination of the following words:

```
ACTG TACG ACGT AATC AGTG TCCA CAAG CCTG
```

This can be depicted graphically as in figure.

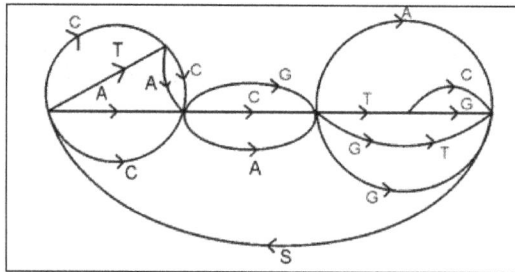

Figure: Graphical Representation of Word Probability Sequence

The various probability (transition and digram) are calculated according to the algorithm given below:

Algorithm

SERIES (DATA, N, LOC): Here DATA is a linear array with N elements and LOC acts as a pointer that keeps the record of occurrence of the nucleotide.

```
1.  Initialize LOC = 0 and initialize all variables to zero.

2.  Repeat while LOC ≠ N,

IF DATA [LOC] = 'A'

Set A = A+ 1
```

```
IF DATA [LOC +1 ] = 'A' SET AA = AA+1

IF DATA [LOC +1 ] = 'C' SET AC = AC+1

IF DATA [LOC +1 ] = 'T' SET AT = AT+1

IF DATA [LOC +1 ] = ' G' SET AG = AG +1

Else IF DATA [LOC] = ' C' Set    C = C+1

IF DATA [ LOC + 1] = 'A' SET CA = CA+1

IF DATA [LOC +1 ] = ' C' SET CC = CC+1

IF DATA [LOC +1 ] = ' T' SET CT = CT+1

IF DATA [LOC +1 ] = ' G' SET CG = CG +1

Else IF DATA [LOC] = 'T'

Set T = T+1

IF [LOC +1 ] = 'A' SET TA = TA+ 1

IF [LOC +1 ] = 'C' SET TC = TC+1

IF [LOC + 1 ] = ' T' SET TT = TT + 1

IF [LOC + 1 ] = ' G' SET TG = TG +1

Else IF DATA [LOC ] = ' G'

Set G = G+1

IF DATA [LOC +1 ] = 'A' SET GA = GA +1

IF DATA [LOC +1] = 'C' SET GC = GC +1

IF DATA [LOC +1] = 'T' SET GT = GT +1 IF DATA [LOC+1] = 'G' SET GG = GG+1

   3. SET A' = A/N, C' = C/N, G' = G/N, T' = T / N

AA' = AA* A'; AC' = AC*A'; AT ' = AT* A';

AG ' =AG* A'; CA' = CC* C' ; CC' = CC*C';

CT' = CT * C'; CG' = CG * C'; TA' = TA * T'; TC' = TC * T'; TT' =TT * T'; TG' = TG* T';

GA' = GA *G'; GC' = GC *G'; GT' = GT *

G'; GG' = GG * G';

   4. End
```

Letter Frequency

(i)	P(i)
A	No of A'S / Size of string = (A' Say)
C	No of C'S / Size of string = (C' Say)
T	No of T'S / Size of string = (T' Say)
G	No of G'S / Size of string = (G' say)

Digram Probability

P(i, j)	A	C	T	G
A	No. of AA * A' = AA'	No. of AC * A' = AC'	No. of AT * A' = AT'	No. of AG * A' = AG'
C	No. of CA * C' = CA'	No. of CC * C' = CC'	No. of CT * C' = CT'	No. of CG * C' = CG'
T	No. of TA * T' = TA'	No. of TC * C' = TC'	No. of TT * T' = TT'	No. of TG * T' = TG'
G	No. of GA * G' = GA'	No. of GC * C' = GC'	No. of GT * T' = GT'	No. of GG * G' = GG'

Transition Probability

Pi(j)	A	C	T	G
A	A' * AA'	A' * AC'	A' * AT'	A' * AG'
C	C' * CA'	C' * CC'	C' * CT'	C' * CG'
T	T' * TA'	T' * TC'	T' * TT'	T' * TG'
G	G' * GA'	G' * GC'	G' * GT'	G' * GG'

On using the algorithm on the different H1N1 viruses we get the following transition probability tables:

Type 1: Table

	A	C	T	G
A	0.13	0.06	0.09	0.08
C	0.09	0.04	0.05	0.02
T	0.06	0.05	0.05	0.07
G	0.08	0.04	0.05	0.06

Type 2: Table

	A	C	T	G
A	0.13	0.06	0.09	0.08
C	0.09	0.04	0.04	0.02
T	0.06	0.05	0.06	0.07
G	0.08	0.04	0.04	0.06

Type 3: Table

	A	C	T	G
A	0. 08	0.05	0.08	0.08
C	0.08	0.03	0.03	0.02
T	0.05	0.06	0.06	0.09
G	0.08	0.06	0.06	0.07

A graphical representation for the type 1 H1N1 virus can be constructed based on the digram probability as follows.

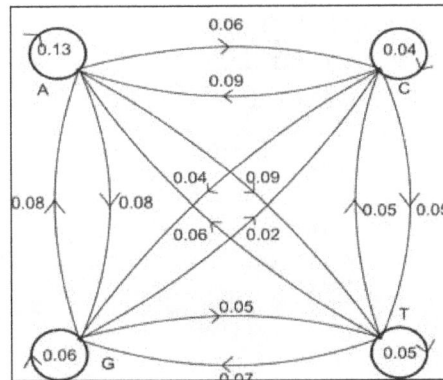

Figure: Graphical Representation of Type 1 H1N1 Sequence

Now if any new sequence comes which has to be categorized as any one of the three types we can use the transition probability table of the existing types and calculate the probability measure. For example consider a new sequence tggtgctggc:

Probability for Type 1: $0.07 * 0.06 * 0.05 * 0.07 * 0.04 * 0.05 * 0.07 * 0.06 * 0.04 = 4.9 \times 10^{-12}$

Probability for Type 2: $0.06 * 0.06 * 0.04 * 0.07 * 0.04 * 0.04 * 0.07 * 0.06 * 0.04 = 2.7 \times 10^{-11}$

Probability for Type 3: $0.09 * 0.07 * 0.04 * 0.09 * 0.06 * 0.06 * 0.09 * 0.07 * 0.06 = 3 \times 10^{-11}$

Since the probability measure for type 3 comes out to be the greatest we can conclude that the chance for the new DNA sample to be the type 3 virus is more.

BLAST

- BLAST is one of the most used algorithms in bioinformatics.

- BLAST includes a family of related algorithms capable of aligning a query sequence against all the sequences in a database in a very fast way.

- BLAST will produce local alignments of the query sequence against the sequences in the database.

- BLAST also generates a expect value that is related with the statistical significance of the alignment.

- BLAST can refer to the family of algorithms, to the NCBI BLAST implementation and to the NCBI BLASTserver.

Algorithm

BLAST uses a very fast algorithm based on words (also known as k-tuples or kmers).

BLAST is many orders of magnitude faster than any Smith & Waterman implementation. To achive that speed it will create less acurate alignments and it will be somewhat less sentitive. The Smith & Waterman algorithm is guaranteed to get always the best alignment, the best scoring one, but it will be much, much slower.

BLAST uses an index of the sequence database in which the positions for all words presents in the sequences are stored. Let's suppose that our database includes only one protein sequence: DS-FASPUIVH. If we create the database index with all the three letter words we would have:

```
sequence: DSFASPUIVH

words:    DSF
            SFA
              FAS
                ASP
                  SPU
                    PUI
                      UIV
                        IVH
```

To do the BLAST search the words for the query sequence are generated and those words are looked for in the database index. This process will be very fast because we have very fast indexing algorithms.

The sequence regions in the database in which the query words are found are known as seeds and will be the seed from which the alignments will be build. One drawback of this algorithm is that if the query and the database do not share words, if they do not share regions with exact matches BLAST won't be able to detect any similiarity at all. That could be well the case with very different sequences. If the sequences are different enough to have no words in common BLAST won't be able to find the alignment. This is the main reason why BLAST is less sentitive than Smith & Waterman. But this word indexing approach is much faster.

The word length will influence the BLAST sensitivity. The shorter the word length the more sentitive will be the search, but the slower will be the process.

Once the seeds have been generated BLAST looks for diagonals in the similiarity matrix. From the seeds the alignments are extended. A query sequence and a single database sequence can generate several alignments. BLAST won't force to join all these alignments into one single alignment. The alignment created by BLAST will not be optimal in most cases. Each alignmed is a High Scoring Pair (HSP) in the BLAST nomenclature. A database sequence and the query sequence can have several HSPs.

e-value

Once we have the alignments we should ask: are the alignments statistically significative?

Let's try to align to random sequences:

```
Query= unknown

        (207 letters)
```

unknown Length = 333 Score = 14.2 bits (25), Expect = 2.9 Identities = 4/16 (25%), Positives = 8/16 (50%) Query: 160 PKRYDWTGKNWVYSHD 175 P ++ W +SH+ Sbjct: 146 PTKHAWQEIGKEF-SHE 161.

So, we can obtain an alignmnet between two random sequences. That is even more problematic for the blast case because we are trying to align a query sequence to millions of sequences in a sequence database. In this case we could obtain many alignments just by chance.

One way to check if an alignment that we obtain has been obtained just by chance or if it is really a significative alignment is to compare its score with the scores of alignments between random sequences. Let's supposee that the alignment of random sequences tend to give scores between 0 and 30. If we obtain an alignment between the query sequence and the database of 15 we might think that is just a random fluke, but if we get a score of 70 we might be quite sure that the alignment is truly significative. You expect to get lots of random alignments with low scores, but few with high scores. This is the approach followed by BLAST to assertain the significance of its alignments.

BLAST creates the distribution of scores between random sequences in the database. For every alignment (HSP) it compares its score to the distribution of the random sequence alignments. The e-value (expect) is the number of hits that one can expect to see by chance in a database. The higher the score for an HSP the lower the e-value will be. For example, an E value of 1 assigned to a hit can be interpreted as meaning that in a database of the current size one might expect to see 1 match with a similar score simply by chance.

The BLAST e-value depends of the database queried. We can not compare e-values obtained for different databases.

It is not possible to set a general threshold for e-values, but in real searches it is usual to find that significant alignments have e-values between 1e-5 and 0.

BLAST example:

```
Blast of sequence c15d_01-D10-M13R_c against database arabidopsis

BLASTX 2.2.10 [Oct-19-2004]

Reference: Altschul, Stephen F., Thomas L. Madden, Alejandro A. Schaffer,

Jinghui Zhang, Zheng Zhang, Webb Miller, and David J. Lipman (1997),

"Gapped BLAST and PSI-BLAST: a new generation of protein database search

programs",  Nucleic Acids Res. 25:3389-3402.

Query= c15d_01-D10-M13R_c

        (892 letters)

Database: tair6

        30,690 sequences; 12,653,157 total letters

Searching.................................................done

                                                         Score     E

Sequences producing significant alignments:             (bits) Value
```

```
AT3G28480.1 | Symbol: None | oxidoreductase, 20G-Fe(II) oxygenas...   280   6e-76

AT3G28490.1 | Symbol: None | oxidoreductase, 20G-Fe(II) oxygenas...   258   3e-69

AT5G18900.1 | Symbol: None | oxidoreductase, 20G-Fe(II) oxygenas...   244   4e-65

AT3G06300.1 | Symbol: AT-P4H-2 | Encodes a prolyl-4 hydroxylase ...   234   4e-62

AT1G20270.1 | Symbol: None | oxidoreductase, 20G-Fe(II) oxygenas...   213   1e-55

AT4G35810.1 | Symbol: None | oxidoreductase, 20G-Fe(II) oxygenas...   201   4e-52

AT2G17720.1 | Symbol: None | oxidoreductase, 20G-Fe(II) oxygenas...   192   2e-49

AT5G66060.1 | Symbol: None | oxidoreductase, 20G-Fe(II) oxygenas...   186   2e-47

AT4G35820.1 | Symbol: None | oxidoreductase, 20G-Fe(II) oxygenas...   159   2e-39

AT4G33910.1 | Symbol: None | oxidoreductase, 20G-Fe(II) oxygenas...   149   3e-36

AT2G43080.1 | Symbol: AT-P4H-1 | Encodes a prolyl-4 hydroxylase ...   147   1e-35

AT2G23096.1 | Symbol: None | oxidoreductase, 20G-Fe(II) oxygenas...   137   7e-33

AT4G25600.1 | Symbol: None | ShTK domain-containing protein, sim...   131   6e-31

>AT3G28480.1 | Symbol: None | oxidoreductase, 20G-Fe(II) oxygenase family
         protein, similar to prolyl 4-hydroxylase, alpha subunit,
         from Gallus gallus (GI:212530), Rattus norvegicus
         (GI:474940), Mus musculus (SP:Q60715); contains PF03171
         20G-Fe(II) oxygenase superfamily domain |
         chr3:10677275-10679525 REVERSE | Aliases: MFJ20.16
         Length = 316

 Score =  280 bits (717), Expect = 6e-76

 Identities = 141/235 (60%), Positives = 174/235 (74%), Gaps = 1/235 (0%)

 Frame = +2

Query: 191 NRFPKMLLHNNDMYESVIRMKTGGSAITIDPTRVIQLSSKPRAFLYEGFLSYEECQHLIH 370
            NRF  +  +N   SVI+MKT S+   DPTRV QLS  PR FLYEGFLS EEC H I
Sbjct: 25  NRF--LTRSSNTRDGSVIKMKTSASSFGFDPTRVTQLSWTPRVFLYEGFLSDEECDHFIK 82

Query: 371 LAKGKLRQSLVAAG-TGESVASKERTSTGMFLRKAQGKIVARIESRIAAWTFLPLDNGEP 547
            LAKGKL +S+VA   +GESV S+ RTS+GMFL K Q  IV+ +E+++AAWTFLP +NGE
Sbjct: 83  LAKGKLEKSMVADNDSGESVESEVRTSSGMFLSKRQDDIVSNVEAKLAAWTFLPEENGES 142

Query: 548 IQILRYENGQKYEPHFDFFQDPGNIAIGGHRIATILMYLSDVEKGGETVFPNSPVKLSEE 727
            +QIL YENGQKYEPHFD+F D N+ +GGHRIAT+LMYLS+VEKGGETVFP   K ++
Sbjct: 143 MQILHYENGQKYEPHFDYFHDQANLELGGHRIATVLMYLSNVEKGGETVFPMWKGKATQL 202

Query: 728 EKGDLSECAXVGYGVRPKLGDALLFFSMNPNVTPDATSYHGSCPVIEGEKMVCTK 892
            +ECA  GY V+P+ GDALLFF+++PN T D+ S HGSCPV+EGEK   T+
```

```
Sbjct: 203 KDDSWTECAKQGYAVKPRKGDALLFFNLHPNATTDSNSLHGSCPVVEGEKWSATR 257
```

>AT3G28490.1 | Symbol: None | oxidoreductase, 2OG-Fe(II) oxygenase family
          protein, similar to prolyl 4-hydroxylase, alpha subunit,
          from Caenorhabditis elegans (GI:607947), Mus musculus
          (SP:Q60715), Homo sapiens (GI:18073925); contains
          PF03171 2OG-Fe(II) oxygenase superfamily domain |
          chr3:10680286-10681891 REVERSE | Aliases: MFJ20.17
          Length = 288

 Score = 258 bits (659), Expect = 3e-69

 Identities = 128/211 (60%), Positives = 159/211 (75%), Gaps = 2/211 (0%)

 Frame = +2

```
Query: 266 AITIDPTRVIQLSSKPRAFLYEGFLSYEECQHLIHLAKGKLRQSLVAAG--TGESVASKE 439
           ++DPTR+ QLS  PRAFLY+GFLS EEC HLI LAKGKL +S+V A   +GES  S+
Sbjct: 24  SFSVDPTRITQLSWTPRAFLYKGFLSDEECDHLIKLAKGKLEKSMVVADVDSGESEDSEV 83

Query: 440 RTSTGMFLRKAQGKIVARIESRIAAWTFLPLDNGEPIQILRYENGQKYEPHFDFFQDPGN 619
           RTS+GMFL K Q  IVA +E+++AAWTFLP +NGE +QIL YENGQKY+PHFD+F  D
Sbjct: 84  RTSSGMFLTKRQDDIVANVEAKLAAWTFLPEENGEALQILHYENGQKYDPHFDYFYDKKA 143

Query: 620 IAIGGHRIATILMYLSDVEKGGETVFPNSPVKLSEEEKGDLSECAXVGYGVRPKLGDALL 799
           +GGHRIAT+LMYLS+V KGGETVFPN   K + +    S+CA  GY V+P+ GDALL
Sbjct: 144 LELGGHRIATVLMYLSNVTKGGETVFPNWKGKTPQLKDDSWSKCAKQGYAVKPRKGDALL 203

Query: 800 FFSMNPNVTPDATSYHGSCPVIEGEKMVCTK 892
           FF+++ N T D  S HGSCPVIEGEK   T+
Sbjct: 204 FFNLHLNGTTDPNSLHGSCPVIEGEKWSATR 234
```

BLAST Programs

BLAST has different programs to compare different kinds of sequences:

- BLASTN compares a nucleotide sequence with a nucleotide database.

- BLASTP compares a protein sequence with a protein database.

- BLASTX compares the six translated frames of a nucleotide sequence with a protein database.

- TBLASTN compares a protein sequence with the six frame translation of a nucleotide database.

- TBLASTX compares a nucleotide sequence with a nucleotide database, but first it translates both.

Megablast

Megablast is an algorithm derived from BLAST. It is faster than BLAST although a little bit less sensitive. It will be faster than BLAST, but it will miss the match if the sequences are very different.

BLAST 2 Sequences

It uses the BLAST algorithm to do a pairwise alignment. It does not uses a database.

It is a very fast way to get a rough alignment (not as good as a Smith and Waterman one). It has the advantage of trying to align both strands.

PSI-Blast and Delta-Blast

PSI-BLAST and DELTA-BLAST are two alternative algorithms to blastp that are more sensitive. They can be used to find distant relatives in the sequences.

PSI-BLAST (Position-Specific Iterated BLAST) is based on the use of Position-Specific Scoring Matrices(PSSM). A PSSM summarizes the patterns found in several sequences. It calculates the frequency for each nucleotide or aminoacid in every position of a multiple alignment.

```
seq1   A C T G   A T   G

seq2   A C T A   A C   G

PSSM A 1 0 0 0.5 1 0   0

     C 0 1 0 0   0 0.5 0

     G 0 0 0 0.5 0 0   1

     T 0 0 1 0   0 0.5 0
```

PSSMs describe the relevant features in the sequences and are usually used to describe conserved domains in the proteins or in the DNA.

PSI-BLAST uses an iterative approach. In the first steps it runs a standard blastp search. With the significant results it creates a mutiple alignments and from the alignment it calculates a PSSM. In the second iteration PSI-BLAST is able to search the database by aligning the PSSM with the database sequences. The sequences found in this new search are used to create a new PSSM and the process is repeated again.

DELTA-BLAST is similar to PSI-BLAST but in the first search it uses a Conserved Domain Database. With the results creates a PSSM and uses that to search in the protein database.

Needleman–Wunsch Algorithm

Needleman–Wunsch Algorithm was developed by Saul B. Needleman and Christian D. Wunsch in 1970. It is one of the most basic and important algorithm of bioinformatics It was designed to compare biological sequences and was one of the first applications of dynamic programming to the biological sequence comparison. This algorithm is usually used for global alignment of two sequences (nucleotide or amino acids).

For the purpose of explanation, first summarizing the algorithm in five steps and then I will move forwards with examples and explanations.

- Consider all the possible pairs of residues from two sequences, the best way is to generate a 2D matrix of two sequences. We will need 2 such matrices one for sequence and one for scores.

- Initialize the score matrix: Scoring matrices are used to determine the relative score made by matching two characters in a sequence alignment. There are many flavors of scoring matrices for amino acid sequences, nucleotide sequences, and codon sequences, and each is derived from the alignment of "known" homologous sequences. These alignments are then used to determine the likelihood of one character being at the same position in the sequence as another character.

- Gap penalty: Usually there are high chances of insertions and deletions (indels) in biological sequences but one large indels is more likely rather than multiple small indels in a given sequences. In order to tackle this issue we give two kind of penalities ; Gap opening penalty (relatively higher) and gap extension penalty (relatively lower).

- Calculate scores and fill the traceback matrix.

- Deduce the alignment from the traceback matrix.

Now, let's with a small example, the alignment of two sequences using BLOSUM62 substitution matrix. Assume sequence A as "SEND" and sequence B as "AND" and gap opening penalty of 10 (no gap extension):

- The Score And Traceback Matrices: The cells of the score matrix are labelled C(i; j) where i = 1; 2; ::::; N and j = 1; 2; ::::; M.

		S	E	N	D		S	E	N	D
	0	−10	−20	−30	−40	done	left	left	left	left
A	−10					up				
N	−20					up				
D	−30					up				

- Scoring: The score matrix cells are filled by row starting from the cell C(2; 2). The score of any cell C(i; j) is the maximum of:

1. $q^{diag} = C(i\ 1; j\ 1) + S(i; j)$

2. $q^{up} = C(i\ 1; j) + g$

3. $q^{left} = C(i; j\ 1) + g$

where $S(i; j)$ is the substitution score for letters i and j, and g is the gap penalty. The value of the cell $C(i; j)$ depends only on the values of the immediately adjacent northwest diagonal, up, and left cells.

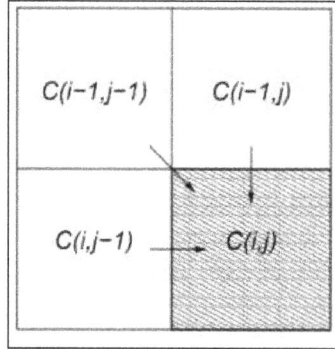

The Needleman-Wunsch Progression

1.   The first step is to calculate the value of $C(2; 2)$.

		S	E	N	D			S	E	N	D
	0	-10	-20	-30	-40		done	left	left	left	left
A	-10	?					up	?			
N	-20						up				
D	-30						up				

- $q^{diag} = C(1; 1) + S(S; A) = 0 + 1 = 1$

- $q^{up} = C(1; 2) + g = 10 + (10) = 20$

- $q^{left} = C(2; 1) + g = 10 + (10) = 20$

Where $C(1; 1)$, $C(1; 2)$, and $C(2; 1)$ are read from the score matrix, and $S(S; A)$ is the score for the S <-> A taken from the BLOSUM62 matrix. For the score matrix $C(2; 2) = q^{diag}$ which is 1. The corresponding cell of the traceback matrix is "diag":

		S	E	N	D			S	E	N	D
	0	-10	-20	-30	-40		done	left	left	left	left
A	-10	1					up	diag			
N	-20						up				
D	-30						up				

2. The next step is to calculate C(2; 3):

- q^{diag}= C(1; 2) + S(E; A) = 10 + 1 = 11

- q^{up}= C(1; 3) + g = 20 + (10) = 30

- q^{left} =C(2; 2) + g = 1 + (10) = 9

Thus C(2; 3) = 9 and the corresponding cell of the traceback matrix is "left". After complete calculations of all cells, the score and traceback matrices are:

		S	E	N	D			S	E	N	D
	0	-10	-20	-30	-40		done	left	left	left	left
A	-10	1	-9	-19	-29		up	diag	left	left	left
N	-20	-9	-1	-3	-13		up	diag	diag	diag	left
D	-30	-19	-11	2	3		up	up	diag	diag	diag

- The Traceback: Traceback is the process of deduction of the best alignment from the traceback matrix. The traceback always begins with the last cell to be filled with the score, i.e. the bottom right cell. One moves according to the traceback value written in the cell. There are three possible moves: diagonally (toward the top-left corner of the matrix), up, or left. The traceback is completed when the first, top-left cell of the matrix is reached ("done" cell). The traceback performed on the completed traceback matrix:

		S	E	N	D
	done	left	left	left	left
A	up	diag ← left	left	left	
N	up	diag	diag	diag	left
D	up	up	diag	diag	diag

Traceback starts here

- The Best Alignment: The alignment is deduced from the values of cells along the traceback path, by taking into account the values of the cell in the traceback matrix:

1. Diag - the letters from two sequences are aligned

2. Left - a gap is introduced in the left sequence

3. Up - a gap is introduced in the top sequence

Sequences are aligned backwards.

The Traceback Step-By-Step

1. The first cell from the traceback path is "diag" implying that the corresponding letters are aligned:

    D

    D

2. The second cell from the traceback path is also "diag" implying that the corresponding letters are aligned:

    ND

    ND

3. The third cell from the traceback path is "left" implying the gap in the left sequence (i.e. we stay on the letter A from the left sequence):

    END

    -ND

4. The fourth cell from the traceback path is also "diag" implying that the corresponding letters are aligned. We consider the letter A again, this time it is aligned with S:

    SEND

    A-ND

Compare with The Exhaustive Search:

1. The best alignment via the Needleman-Wunsch algorithm:

    SEND

    A-ND

2. The exhaustive search:

    SEND

    -AND score: +1

    A-ND score: +3 the best

    AN-D score: -3

    AND- score: -8

Pros and Cons: In this example we considered very short sequences SEND and AND which are much easier to compare using exhaustive search but if we think of real problems that contains longer sequences the situation quickly turns against it. For example; two 12 residue sequences would require considering ~ 1 million alignments and two 150 residue sequences would require considering ~ 10^88 alignments (~ 10^78 is the estimated number of atoms in the Universe). On the other hand for two 150 residue sequences the Needleman-Wunsch algorithm requires only filling a 150 * 150 matrix, which is not much computationally expensive. The major problem is Needleman-Wunsch algorithm works in the same way regardless of the length or complexity of sequences which provides heuristic algorithm an upper hand over this algorithm but Needleman-Wunsch algorithm guarantees to provide the best alignment.

References

- Artificial-neural-network: data-flair.training, Retrieved 28 February, 2019

- Genetic-algorithm-explanation-and-perl-code: bioinformaticsreview.com, Retrieved 1 March, 2019

- 11-smith-waterman-algorithm: bioinfoguide.com, Retrieved 27 January, 2019

- Blast, biotech: comav.upv.es, Retrieved 6 June, 2019

- 10-needleman-wunsch-algorithm: bioinfoguide.com, Retrieved 10 April, 2019

Chapter 3

Branches and Applications of Bioinformatics

There are several branches of bioinformatics such as translational bioinformatics, integrative bioinformatics and structural bioinformatics. The topics elaborated in this chapter will help in gaining a better perspective about these branches of bioinformatics and their applications.

Integrative Bioinformatics

Integrative Bioinformatics deals with the development of methods and tools to solve biological problems as well as providing a better understanding or new knowledge about biochemical phenomena by means of data integration and computational experiments

High blood pressure or hypertension is an established risk factor for a myriad of cardiovascular diseases. Genome-wide association studies have successfully found over nine hundred loci that contribute to blood pressure. However, the mechanisms through which these loci contribute to disease are still relatively undetermined as less than 10% of hypertension-associated variants are located in coding regions. Phenotypic cell-type specificity analyses and expression quantitative trait loci show predominant vascular and cardiac tissue involvement for blood pressure-associated variants. Maps of chromosomal conformation and expression quantitative trait loci (eQTL) in critical tissues identified 2,424 genes interacting with blood pressure-associated loci, of which 517 are druggable. Integrating genome, regulome and transcriptome information in relevant cell-types could help to functionally annotate blood pressure associated loci and identify drug targets.

Elevated blood pressure (BP) or hypertension is a heritable chronic disorder, considered the single largest contributing risk factor in disease burden and premature mortality. High systolic and/or diastolic BP reflects a higher risk of cardiovascular diseases. Genome-wide association studies (GWAS) have found association of 905 loci to BP traits (systolic - SBP, diastolic - DBP and pulse pressure -PP) to date. The use of larger sample sizes has helped to identify additional variants, as demonstrated by the most recent study including over 1 million people that has identified 535 novel BP loci. Still, this collective effort thus far has not entirely elucidated the complete genetic contribution to BP, estimated to be approximately 50–60%.

To add to this complexity, 90.7% of the 905 BP-associated index variants are located in intronic or intergenic regions. Causal variants are also difficult to pinpoint because of linkage disequilibrium (LD). There is now vast evidence that non-coding variants associated with disease interrupt the action of regulatory elements crucial in relevant tissues for that particular disease. BP loci are not only linked to cardiovascular disease but also to other diseases, suggesting that BP-associated variants can result in a wide range of phenotypes. Tissue specificity of genetic loci may be relevant for organ specific disease progression. For example, variants altering expression in heart may more likely affect disease progression through heart-mediated processes rather than kidney-mediated processes, and some patients may suffer of left ventricular hypertrophy while others may develop nephropathy. Thus, investigating the influence of BP variants in critical cell-types is essential in understanding disease risk and biology, and assessing the possible translation of an associated locus into a drug target. The public availability of regulatory annotations in several tissues by projects such as ENCODE, Roadmap and GTEx has enabled integration of epigenetic modifications, expression quantitative trait loci (eQTLs) and –omics information with GWAS data. Integrative approaches are useful for prioritizing genes from known GWAS loci for functional follow-up, detecting novel gene-trait associations, inferring the directions of associations, and potential druggability.

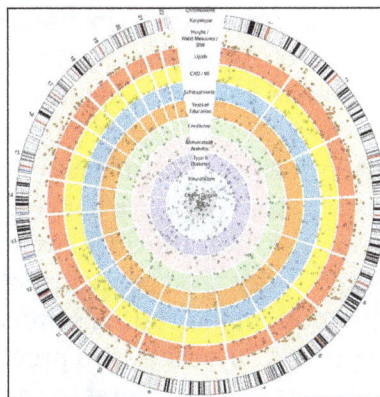

Figure: Circos plot showing the 10 traits from the GWAS catalog (37) with the largest number of loci also associated to BP, as identified by PhenoScanner (38) at p < 0.05 (Supplemental Methods). The outer ring represents the genomic/chromosomal location (hg19).

The following inner rings show the associations to different traits. Beige: body measurements (height, body mass index (BMI), weight, waist/hip ratio, hip circumference, waist circumference. N = 358). Red: lipids (high-density lipoprotein (HDL), low-density lipoprotein (LDL), triglycerides, total cholesterol. N = 226). Yellow: coronary artery disease (CAD)/myocardial infarction (MI) (N = 206). Blue: schizophrenia (N = 135). Orange: years of education attendance (N = 101). Light green: creatinine (N = 88). Light pink: rheumatoid arthritis (N = 78). Purple: type II diabetes (N = 73). Light turquoise: neuroticism (N = 69). Light grey: Crohn's disease (N = 67).

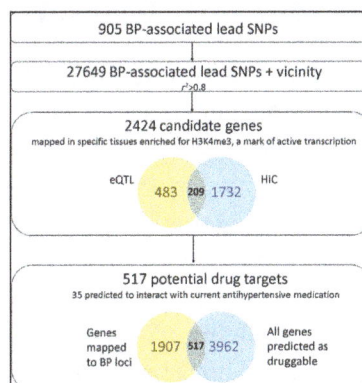

Figure: Diagram of analytical steps that can be followed for variant prioritization and translation of association to a potential drug target. Each step is accompanied by examples of publicly available data (green boxes on the left) and tools (yellow boxes on the right) that can be used.

Integrative Approaches using – Omics Data

Remarkable advances have been made recently towards a better comprehension of BP genetics, the biology of disease and translation towards new therapeutics, boosted by the widespread application of high-throughput genotyping technologies. At the same time, most BP-associated variants are non-coding, making the conversion of statistical associations into target genes a great challenge. SIFT, PROVEAN, PolyPhen, CONDEL and more recently CADD are scoring algorithms developed for predicting the effect of amino acid changes. Only 98 out of the 905 lead BP-associated SNPs reflect a CADD score above 12.37, a threshold suggested by Kicher et al. as deleterious. However, the causal variant inside the locus might reflect a different CADD score than the lead SNP, and pinpointing the mechanisms disturbed by the variation remains a challenge.

New strategies that make use of regulatory annotations in disease-relevant tissues have greatly expanded our ability to investigate the processes involved in BP. In particular, annotation of histone modifications and regions of open chromatin allow the identification of active transcription in specific-cell types. Similarly, maps of DNA variants affecting expression in a cell-type specific manner will be integral in BP loci interpretation. A list of cardiovascular-related cell-types researched by the ENCODE Project is presented by Munroe et al. Such data can be integrated with GWAS results using bioinformatics tools. For instance, FUMA provides extensive functional annotation for all SNPs in associated loci and annotates the identified genes in biological context. FunciSNP investigates functional SNPs in regulatory regions of interest. Ensemble's Variant Effect Predictor (VEP) determines the effect of variants on genes, transcripts, and protein sequence, as well as regulatory regions, also outputting SIFT, Polyphen and CADD scores for each variant, among other information. Although such integrative tools are useful for variant prioritization and interpretation, not all take into consideration tissue specificity aspects. RegulomeDB, for example, is a database that annotates SNPs with known and predicted regulatory elements in the intergenic regions of the human genome, calculating a score that reflects its evidence for regulatory potential. However, the scoring procedure can only be performed across all available tissue types. In addition, several databases containing a broad range of tissues were made publicly available since the last update of RegulomeDB, that could be included in the tool. Together, these resources have been useful in prioritizing genes and variants in associated loci for functional follow-up experiments in many post-GWAS analyses, and can be implemented in interpretation of BP-associated loci.

Transcription Regulation: Histone Modifications and Open Chromatin

As genomic coordinates of active regulatory elements may be mapped using unique functions of chromatin, the characterization of chromatin changes in the genome in specific cell-types can be used to identify DNA variants disturbing active regulatory elements. The four core chromatin histones, H_2A, H_2B, H_3 and H_4, can suffer posttranslational modifications, such as acetylation or methylation. These histone modifications indicate active (euchromatin) or repressed (heterochromatin) chromatin structure, defining regulation and gene transcription. Acetylation of histones H_3 and H_4, and H_3 methylation at Lys_4 ($H_3K_4me_3$), for instance, correlate with gene transcription, whereas methylation at Lys_9 correlates with gene silencing. These modifications provide a robust readout of active regulatory positions in the genome, and have been employed for annotation in several studies. Histone modifications influencing arterial pressure have been observed in many tissues, including vascular smooth muscle. An updated phenotypic cell-type specificity analysis of the 905 BP loci using $H_3K_4me_3$ mark in 125 tissues. The most significant cell-types are cardiovascular-related. Other tissues with high rank in specificity are smooth muscle, fetal adrenal gland, embryonic kidney cells, CD34 and stem-cell derived CD56 +mesoderm cultured cells. These results are consistent with anal-yses using DNase I hypersensitivity sites (DHSs), which indicate likely binding sites of transcription factors. These results add more evidence that BP loci are enriched on regions of open chromatin, regulating transcription in a broad range of tissues.

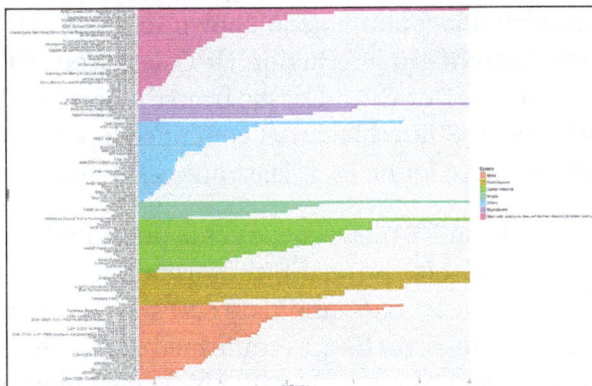

Figure: Ranked tissues after phenotypic cell-type specificity analysis of 905 BP SNPs using 125 $H_3K_4me_3$ datasets on human tissue

Methylation

In addition to histone modifications that promote transcription, BP loci have also been studied for their enrichment on DNA methylation, known to have the opposite regulatory effect. The methyl-ation of CpG sites, presented by CpG islands in promoters, affects binding of transcription factors, resulting in gene silencing. Abnormal CpG methylation is found in hypertension, and in many other complex diseases. Recently, Kato et al. identified a ~2 fold enrichment associating BP vari-ants and local DNA methylation. The study also demonstrates that DNA methylation in blood cor-relates with methylation in several other tissues. These observations add to previous indications on the function of DNA methylation in regulating BP.

Measuring the Impact of BP Risk Alleles on Gene Expression: eQTLs

Expression quantitative trait loci (eQTL) are regions harbouring nucleotides correlating with alter-ations in gene expression. Linking transcription levels to complex traits has been a follow-up step adopted by many studies, driven by the increase in available data of expression patterns across tissues and populations. Warren et al. found that 55.1% of their identified BP-associated loci have SNPs with eQTLs in at least one tissue from GTex repository, with arterial tissue most frequently observed (29.9% of loci had eQTL in aorta and/or tibial artery). A great enrichment of eQTLs in artery was also observed by Evangelou et al., who identified 92 novel loci with eQTL enrichment in arterial tissue and 48 in adrenal tissue. In summary, these studies also suggest that BP loci exert a regulatory effect mostly in vascular and cardiac tissues.

Finding the Targets: Chromosome Confirmation Capture Techniques

Mapping variation to target genes is one of the greatest challenges in the post-GWAS era, and different strategies have been developed to this end. One approach is the use of chromosome confirmation capture [3C, 4C, Hi-C]. These techniques capture chromosome interactions, resulting in networks of interacting genetic loci.

Warren et al. made use of this resource to investigate the target genes of non-coding SNPs, using Hi-C data from endothelial cells (HUVECs). Distal potential genes were found on 21 loci, and these genes were enriched for regulators of cardiac hypertrophy in pathway analysis. Kraja et al. also explored long-range chromatin interactions using endothelial precursor cell Hi-C data, finding the link between an associated loci and a gene known to affect cell growth and death. More recently, Evangelou et al. used chromatin interaction Hi-C data from HUVECs, neural progenitor cells (NPC), mesenchymal stem cells (MSC) and tissue from the aorta and adrenal gland to identify distal affected genes. They found 498 novel loci that contained a potential regulatory SNP, and in 484 loci long-range interactions were found in at least one cell-type.

A list of human HiC data available on BP relevant tissues is presented. An updated version of variant to gene mapping making use of this chromatin conformation data. Promoter regions of 1,941 genes were found to interact with the 27,649 candidate SNPs (905 BP associated SNPs and vicinity). Integration with eQTL data on relevant tissues confirmed 209 of the genes mapped, and added additional 483 genes. One main goal of understanding biological mechanisms of GWAS associations and affected genes is to be able to therapeutically target them. Assessment of the druggability of a BP-associated locus depends on several factors, but overlap of these results with a recent effort on druggability suggests that 517 of these 2,424 genes are druggable, and 35 mapped genes are also predicted to interact with common drugs for treatment of hypertension. Interestingly, 1,774 of the genes mapped are physically located outside BP-associated loci. These results support the hypothesis that BP GWAS loci act on tissue specific regulatory gene networks. Importantly, they also show that the use of long range chromatin interaction maps can reliably identify target genes even outside the risk locus.

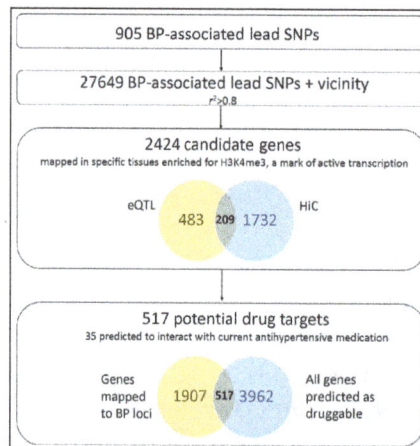

Figure: Diagram illustrating the results of our integrative approach.

Translational Bioinformatics

Translational Bioinformatics (TBI) has become a key component of biomedical research in the era of precision medicine. Development of high-throughput technologies and electronic health records has caused a paradigm shift in both healthcare and biomedical research. Novel tools and methods are required to convert increasingly voluminous datasets into information and actionable knowledge.

According to the American Medical Informatics Association (AMIA), translational bioinformatics (hereafter "TBI") is "the development of storage, analytic, and interpretive methods to optimize the transformation of increasingly voluminous biomedical data, and genomic data, into proactive, predictive, preventive, and participatory health" Put more simply, it is the development of methods to transform massive amounts of data into health. Dr. Russ Altman from Stanford University delivers a year-in-review talk at AMIA's summit on TBI. In his 2014 presentation he provided the following definition for TBI: "informatics methods that link biological entities (genes, proteins, and small molecules) to clinical entities (diseases, symptoms, and drugs)—or vice versa" (gives a visual depiction of the way in which TBI fits within the bigger picture of biomedical informatics and transforming data into knowledge . Along the X axis is the translational spectrum of bench-to-bedside, while the Y axis from top to bottom represents the central dogma of informatics, transforming data to information and information to knowledge. Toward the discovery end of the spectrum (the bench) is bioinformatics, which includes storage, management, analysis, retrieval, and visualization of biological data, often in model systems. The discovery end of the spectrum has some overlap with computational biology, particularly in the context of systems biology methods. Toward the clinical end of the spectrum (bedside) is health informatics. TBI fits in the middle of this space. On the data-to-knowledge spectrum, data collection and storage are the beginning steps. After that comes data processing, analysis, and then interpretation, thereby transforming the information that has been gleaned from the data into actual knowledge, useful in the context of clinical care, or for further research. In that way the data go from just being "bits"—1's and 0's—to new knowledge and actionable insights.

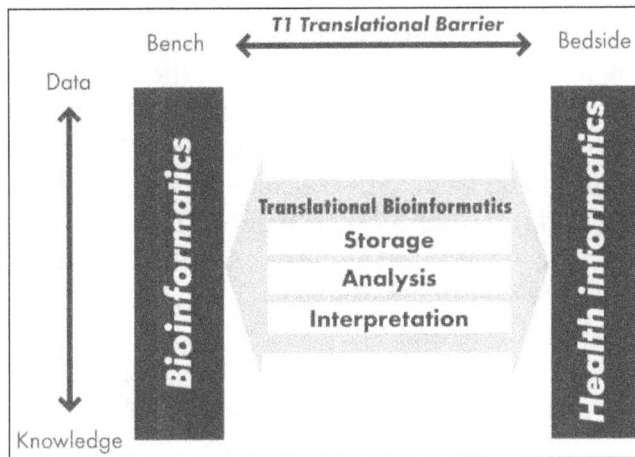

Figure: Translational Bioinformatics in context

The Y axis depicts the "central dogma" of informatics, converting data to information and information to knowledge. Along the X axis is the translational spectrum from bench to bedside. Translational bioinformatics spans the data to knowledge spectrum, and bridges the gap between bench research and application to human health. The figure was reproduced from with permission from Springer.

TBI as a field has a relatively short history. In the year 2000, the initial drafts of the human genome were released, arguably necessitating this new field of study. In 2002, AMIA held its annual symposium with the name "Bio*medical Informatics: One Discipline", meant to recognize and emphasize the spectrum of subdisciplines. In 2006, the term itself was actually coined by Atul Butte and Rong Chen at the AMIA annual symposium in a paper entitled "Finding disease-related genomic experiments within an international repository: first steps in translational bioinformatics" . In 2008, AMIA had its first annual AMIA summit on TBI, chaired by Dr. Butte. Year 2011 saw the first annual TBI Conference in Asia, held in Seoul, Korea. Finally, an online textbook on TBI was published in 2012 by PLoS Computational Biology, edited by Maricel Kann. Initially intended to be a traditional print textbook, this resource was published using an open source model, making it freely available on the Internet.

TBI Today

A study categorized recent themes in the field of TBI into four major categorizations: (1) clinical "big data", or the use of electronic health record (EHR) data for discovery (genomic and otherwise); (2) genomics and pharmacogenomics in routine clinical care; (3) omics for drug discovery and repurposing; and (4) personal genomic testing, including a number of ethical, legal, and social issues that arise from such services.

Big Data and Biomedicine

As technology enables us to take an increasingly comprehensive look across the genome, transcriptome, proteome, etc., the resulting datasets are increasingly high-dimensional. This in turn requires a larger number of samples in order to achieve the statistical power needed to detect the true signal. The past decade or so has seen an increasing number of large-scale bio-repositories intended for clinical and translational research all over the world. These projects comprise both information and biospecimens from individual patients, enabling researchers to reclassify diseases based on underlying molecular pathways, instead of the macroscopic symptoms that have been relied on for centuries in defining disease. These various projects involve different models of participation, ranging from explicit informed consent to use of de-identified biospecimens and their associated clinical information from EHRs (also de-identified). The informed consent is the most ethically rigorous model, but also the most expensive. The use of de-identified specimens and data is more scalable, and financially feasible. However, as complete genomic data are increasingly used, it is impossible to truly de-identify these data. This raises ethical issues regarding patient privacy and data sharing. In the United States, legislation known as the "Common Rule" addresses these issues. In 2015, a notice of proposed rule-making (was released to solicit feedback on some major revisions to the law, which was originally passed in 1991. Much has changed within biomedical research in the intervening years.

In order to accrue the numbers of samples required for the "big data" discipline that biomedical research is becoming, the ability to use patient data and samples in research would be of significant benefit. One major point addressed in the aforementioned NPRM is the ability for patients to give broad consent for future use of data and samples, without knowing the specifics of research studies ahead of time. As we move toward the learning healthcare system (LHS) model in which every encounter is an additional data point, explicit research registries will become less relevant. They will be too expensive to maintain, and larger numbers of patients/participants will be available through federated initiatives that allow a researcher to query across institutions regionally, nationally, and even internationally. The National Patient-Centered Clinical Research Network (PCORnet) takes this approach, enabling clinical outcome research through federated pragmatic clinical trials. Importantly, this initiative emphasizes partnership with patients and their advocates, so that they are empowered as collaborators, with a say in what research questions matter most.

LHS is about moving from evidence-based practice, i.e., clinical care decisions based on conscientious use of current best evidence, to practice-based evidence, i.e., the generation of evidence through collection of data in the real-world as opposed to the artificially-controlled environment of randomized clinical trials. In recent decades, the biomedical enterprise has strived to practice medicine in a way that is supported by the best possible evidence from randomized clinical trials. But clinical trials have their own issues. They are expensive, and they tend to be very different from real life scenarios. Criteria for inclusion in a trial often include the absence of common comorbidities or use of common medications. Compliance tends to be high, but the cohort being studied is often not representative of the target population for the treatment in question. In LHS, translation becomes bi-directional. Research is used to inform practice, whereas data that are generated in the course of clinical care can in turn be used for both hypothesis generation and validation through pragmatic trials. Data derived from clinical care can thus inform clinical guidelines and future practice.

Secondary Use

Secondary use of data refers to data that are created or collected through clinical care. In addition to use in caring for the patient, these data may also be crucial for operations, quality improvement, and comparative effectiveness research. Some assert that the term "secondary use" should give way to the term "continuous use." They argue against the notion that data collected at the point of care are solely for clinical use, and everything else is secondary. We should be maximally leveraging this valuable information. Nonetheless, there is a legitimate concern about data quality. Data in the EHR are often sparse, incomplete, even inaccurate. This makes these data wholly unsuitable for certain purposes, but still sufficient for others. For instance, Frankovich et al. described a case in which an adolescent lupus patient was admitted with a number of complicating factors that put her at risk for thrombosis. The medical team considered anti-coagulation, but were concerned about the patient's risk of bleeding. No guideline was available for this specific case, and a survey of colleagues was inconclusive. Through the institution's electronic medical record data warehouse, Frankovich and colleagues were able to look at an "electronic cohort" of pediatric lupus patients who had been seen over a 5-year period. Of the 98 patients in the cohort, 10 patients had developed clots, with higher prevalence in patients with similar complications as the patient in question. Using this real-time analysis based on evidence generated in the course of clinical care, Frankovich and colleagues were able to make an evidence-based decision to administer anti-coagulants. Subsequently, researchers at Stanford University have proposed a "Green Button" approach to formalize this model of real-time decision support derived from aggregate patient data and data capture to help inform future research and clinical decisions.

TBI tends to focus on molecules, newly accessible in high dimensions based on novel high-throughput technologies. Phenotyping is a closely-related challenge, more complex than it might seem. Disease is not binary: even within a very specific type of cancer, a tumor's genomic profile may be quite different among the precise sampling locations and sizes. There are a number of groups focusing on this problem: the Electronic Medical Records and Genomics (eMERGE) Network, the NIH Collaboratory, PCORnet, and the MURDOCK Study, among others. The Phenotype KnowledgeBase website is a knowledge base of phenotypes, offering a collaborative environment to build and validate phenotype definitions. The phenotypes are not (yet) computable, but it serves as a resource for defining patient cohorts in specific disease areas. Richesson et al. looked at type 2 diabetes, a phenotype that one might expect to be fairly straightforward. But defining type 2 diabetes mellitus (T2DM) using the International Classification of Disease version 9 (ICD9) codes, diabetes-related medications, the presence of abnormal labs, or a combination of those factors resulted in very different counts for the number of people diagnosed with T2DM in Duke's data warehouse. Using only ICD9 codes gave 18,980 patients, while using medications yielded 11,800. Using ICD9 codes, medications, and labs all together yielded 9441 patients. Note that the issue is not just a matter of semantics and terminology, where if everyone could agree to a single definition and use the same code, then the terms would become uniform. For different purposes, different definitions of diabetes may be needed, depending on whether the use case involves cohort identification or retrospective analysis. In different cases, one might care more about minimizing false positives (e.g., retrospective analysis) or maximizing true positives (e.g., surveillance or prospective recruitment).

Thousands of papers have been published describing genome-wide association studies (GWAS), in which researchers look across the entire genome to find SNPs that are statistically enriched for a given phenotype (usually a disease) compared with healthy controls. Researchers at Vanderbilt University turned this approach on its head, developing a method known as phenome-wide association studies Instead of looking at the entire genome, PheWAS evaluates the association between a set of genetic variants and a wide and diverse range of phenotypes, diagnoses, traits, and/or outcomes. This analytic approach asks, for a given variant, do we see an enrichment of a specific genotype in any of these phenotypes? illustrates results using this approach. In standard GWAS analyses, the different color bands at the bottom represent the different chromosomes. In the case of PheWAS, they are different disease areas, e.g., neurologic, cardiovascular, digestive, and skin. Pendergrass et al. used a PheWAS approach for the detection of pleiotropic effects, where one gene affects multiple different phenotypes. They were able to replicate 52 known associations and 26 closely-related ones. They also found 33 potentially-novel genotype–phenotype associations with pleiotropic effects, for example the GALNT2 SNP that had previously been associated with HDL levels among European Americans. Here they detected an association between GALNT2 and hypertension phenotypes in African Americans, as well as serum calcium levels and coronary heart disease phenotypes in European Americans.

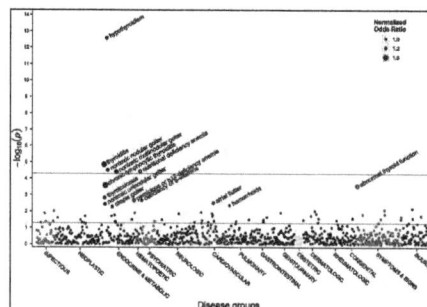

Figure PheWAS Manhattan plot for a given SNP This plot shows the significance of association between SNP rs965513 and 866 different phenotypes. Along the X axis different disease groups are shown in different colors. This is in contrast to an analogous plot for GWAS in which the X axis would represent the different chromosomes. The Y axis reflects the P value for each phenotype. Blue and red horizontal lines represent P value of 0.05 and Bonferroni corrected P value of 5.8×10^{-5}, respectively. PheWAS, phenome-wide association studies; GWAS, genome-wide association studies. The figure was reproduced from with permission from Elsevier.

Another aspect of big data in biomedicine is the use of non-traditional data sources. These were well illustrated, both literally and figuratively, in a 2012 paper by Eric Schadt. A complex and detailed figure showed various data types that could be mined for their effects on human health: weather, air traffic, security, cell phones, and social media among others. But strikingly to those reading the paper just a few years later, the list did not include personal activity trackers, e.g., FitBit, Jawbone, or even the Apple watch. This omission of such a popular technology today is indicative of what a fast-moving field this is.

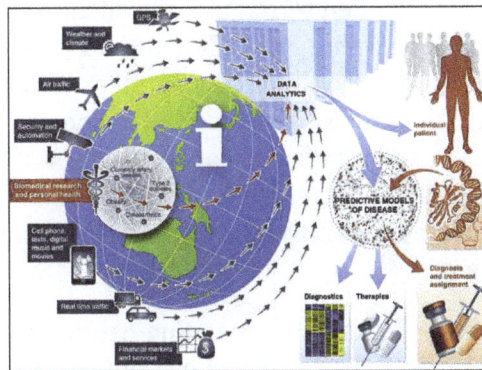

Heterogeneous and non-traditional sources of big data Technological advances have enabled the collection and storage of big data beyond biomedicine, including everything from credit card transactions to security cameras to weather. Notably absent from this 2012 figure are smart watches and fitness tracking devices, which became pervasive in the years that followed.

Genomics in Clinical Care

One sees a number of examples of how genomic data are used in clinical care in the context of pharmacogenomics. But molecular data, and genomic data derived from next-generation sequencing (NGS) in particular, have been used in a number of other contexts as well. One example took place at Stanford's Lucile Packard Children's Hospital, where a newborn presented with a condition known as long QT syndrome. In this specific case, the manifestation was unusually severe—the baby's heart stopped multiple times in the hours after its birth. Long QT syndrome can be caused by mutations in a number of different genes. It is necessary to know which gene harbors the mutation in order to know how to treat the condition. In this case, a whole-genome sequencing (WGS) was performed enabling identification of a previously-studied mutation, as well as a novel copy number variation in the TTN gene that would not otherwise have been detectable through targeted genotyping alone. Moreover, NGS enabled the answer to be obtained in a matter of hours to days instead of weeks.

Another example of DNA sequencing in clinical care involved diagnosis of infectious disease. A 14-year-old boy with severe combined immunodeficiency (SCID) had been admitted to the hospital repeatedly. He had headache, fever, weakness, nausea, and vomiting. His condition continued to decline to the point where he was put into a medically-induced coma. A normal diagnostic work up was "unrevealing" and the doctors were unable to determine the etiology of his condition. The patient was enrolled in a research study for pathogen detection and discovery in hospitalized patients. The protocol for this study involved performing NGS on the subject's spinal fluid. The results included detection of 475 reads corresponding to leptospira infection. Of note, the normal test for leptospirosis involves detection of the patient's antibody response to the infection. In this case, the patient's SCID status prevents such a response, so the infection was not detectable through standard means.

Cancer has been one of the most active areas for the translation of genomic discoveries into changes in clinical care. One of the biggest players in this space is Foundation Medicine, which makes the FoundationOne test, a targeted panel that uses NGS to test all genes known to harbor mutations in solid tumors. Johnson et al. looked retrospectively at approximately 100 patients who had undergone FoundationOne, and found that 83% of them had potentially actionable results from that test, of which 21% received genotype-directed treatment. Explanations for why the indicated treatment was not given in 79% of those cases included the decision to use standard therapy and clinical deterioration. These results indicate that application of genomic technology has transcended the research domain. In many cases, the findings are clinically actionable. Mirroring this fact, it is worth noting that some medical insurance companies in the United States started to offer coverage of the FoundationOne test.

In addition to targeted panels, deep cancer sequencing can help shed light on drug sensitivity and resistance. Wagle et al. describe a case of a 57-year-old woman with anaplastic thyroid cancer. Her specific tumor was initially sensitive to everolimus, a mammalian target of rapamycin (mTOR) inhibitor. Her doctors were able to sequence the tumor before it became resistant, revealing a mutation in TSE2, which encodes a negative regulator of mTOR. Normally mTOR is down-regulated by TSE2, but the mutation caused TSE2 not to down-regulate mTOR to a sufficient level. Therefore, everolimus, which inhibits mTOR, was effective in treating this specific cancer. Later this patient's cancer became drug-resistant, whereupon the newly drug-resistant part of the tumor was sequenced again. It was discovered that an mTOR mutation had cause mTOR not to be inhibited by allosteric inhibitors like everolimus. An allosteric inhibitor binds to the protein in question somewhere other than on the active site of the molecule. Knowing the specific cause for the newly-acquired resistance leads to other treatment options. They were able to switch to mTOR kinase inhibitors, in order to down-regulate the pathway through other mechanisms.

Despite these compelling cases, it is worth noting that at this point, tumor sequencing is the exception in clinical oncology, and not one of the routine procedures. Only after "first line" treatment has failed are tumors sequenced, and even that is largely confined to large academic medical centers. It will be interesting to see if, when, and how that changes.

While blood clotting and cancer are the areas where most actionable pharmacogenomics findings have been made, a notable exception is work by Tang and his colleagues , in which they describe the identification of a genotype-based treatment for T2DM with an alpha2A-adrenergic receptor antagonist. A specific genetic variant causes over-expression of the alpha2A-adrenergic receptor, and impaired insulin secretion. They hypothesized that if they could block the over-expressed receptor, they could increase insulin secretion. Those with the mutation showed a dose response to the drug as measured through levels of insulin. Participants without the mutation showed no such response. This research is especially interesting because it goes beyond cancer and blood thinners, into a chronic disease that is affecting an increasing portion of the world's population.

One of the earliest and best known DNA sequencing success stories was that of Nic Volker, who at age 2 developed severe gastro-intestinal issues resembling Crohn's disease, for which a diagnosis could not be determined. After multiple life-threatening situations and a protracted diagnostic odyssey, Volker's genome was sequenced to look for a causal mutation. Doing so enabled the discovery of a mutation in the gene X-linked inhibitor of apoptosis (XIAP). Equally important, there existed a known therapeutic intervention for disorders caused by XIAP mutations. A progenitor cell transplant was performed and the patient's condition improved. Though Nic is not without ongoing health challenges.

There is also reason for caution regarding use of NGS sequencing in clinical care. Dewey et al. performed Illumina-based WGS on 12 different participants. Confirmatory sequencing was formed on 9 of those participants by Complete Genomics Inc. Their findings included the fact that 10%–19% of inherited disease genes were not covered to accepted standards for SNP discovery. For the genotypes that were called, concordance between the two technologies for previously-described single nucleotide variants was 99%–100%. However, for insertion and deletion variants, the concordance rate was only 53%–59%. Approximately 15% were discordant, and approximately 30% of the variants could not be called by one technology or the other. In addition, inter-rater agreement for whether findings should be followed up clinically was only 0.24 ("fair") by Fleiss' kappa metric. Rater agreement was even worse for cardio-metabolic diseases, with the rate at which the two experts agreed on the need for clinical follow-up worse than random in those cases. Notably, the estimated median cost for sequencing and variant interpretation was about US $15,000, plus the price of the computing infrastructure and data storage. This means the cost of interpretation is significantly more than the proverbial US $1000 genome goal, but is also significantly less than the US $100,000 or $1 million some had feared.

Omics for Drug Discovery and Re-purposing

Much has been said about the protracted process involved in getting a drug through the FDA approval pipeline. Estimates are that the process can take on average 12 years between lead identification and FDA approval. This makes the prospect of drug repurposing an appealing one. Drug repurposing refers to taking an existing, FDA-approved compound and using it to treat a disease or condition other than the one for which it was originally intended. In the past, inspiration for this type of "off label use" has been largely serendipitous. For example, Viagra was initially aimed at treating heart disease, and turned out to be useful for erectile dysfunction. By using a pre-approved compound, early phase clinical trials can be avoided, which can save significant time and money.

Computational approaches to drug repurposing may take a number of different forms as described in two recent reviews. One is to look for molecular signatures in disease and compare those to signatures observed in cells, animal models, or people who have been treated with different drugs. If anti-correlated signatures can be identified between diseases and drugs, administration of that drug for that disease may help cure the condition, or at least to alleviate the symptoms. One of the prominent early examples of a computational approach to drug repositioning was the Broad Institute's Connectivity Map (CMap). The authors identified gene expression signatures for disease states and perturbation by small molecules and then compared those signatures. They made the data available as a resource intended to enable the identification of functional connections between drugs, genes, and diseases. Another example is work by Jahchan et al., in which they identified anti-depressants as potential inhibitors of lung cancer. The authors looked at numerous disease and drug profiles and found an anti-correlation between gene expression seen after administration of anti-depressants and the pattern of expression observed in small cell lung cancer. They next transplanted these types of tumors into mice and found that with the drug, those tumors either shrunk or didn't even grow. They also used indigenous tumors in the mouse model for this type of cancer and found that the anti-depressant had promising results on those tumors and for other endocrine tumors as well. They were able to start a clinical trial, much faster than would have been possible by chance or by various other traditional methods, though unfortunately that trial was ultimately terminated for lack of efficacy. A similar approach was used by Sirota et al. to identify the anti-ulcer drug cimetidine as a candidate agent to treat lung adenocarcinoma and validate this off-label usage in vivo using an animal model of the lung cancer.

An alternative to the largely computational methods described above is an experimental approach to drug repositioning. For example, Nygren et al. screened 1600 known compounds against 2 different colon cancer cell lines. They used Connectivity Map data to further evaluate their findings, and identified mebendazole (MBZ) as having potential therapeutic effect in colon cancer. Finally, Zhu et al. mined data from PharmGKB and leveraged the web ontology language (OWL) to perform semantic inference. They were able to identify potential novel uses and adverse effects of approved breast cancer drugs.

Personalized Genomic Testing

The year 2008 saw the founding of several companies that offered direct-to-consumer (DTC) genetic testing, reporting on a variety of genes for both health and recreational purposes. As of 2016, 23andMe was the last major player standing in the United States, with other companies having been acquired and changing their business models.

DTC genetic testing raises a number of interesting ethical, legal, and social issues. For several years, there was an open question as to whether or not these tests should be subject to government regulation. In November 2013, the US FDA ordered 23andMe to stop advertising and offering their health-related information services. The FDA considered these tests to be "medical devices" and as such to require formal testing and FDA approval for each test. In February 2015, it was announced that the FDA had approved 23andMe's application for a test for Bloom syndrome, and in October 2015 it was announced that the company would once again be offering health information in the form of carrier status for 36 genes. Note that a 23andMe customer is able to download his or her raw genomic data and to use information from other websites to interpret the results, including Promethease, Geneticgenie, openSNP, and Interpretome for health-related associations.

Another important question raised by DTC genetic testing include whether the consumers are ready for this information. Traditionally, patients receive troubling health-related information in a face-to-face conversation with their doctor. There is some concern that patients are not competent or well-equipped to receive potentially distressing news through an Internet link . To help mitigate this concern, 23andMe "locks" certain results, making them accessible only if the user clicks through an additional link, indicating they truly want to know.

What about the healthcare providers? Are they ready to incorporate genomic data, patient-supplied or otherwise, into treatment decisions? In a case described in 2012, a 35-year-old woman informed her fertility care provider that her 23andMe results revealed a relatively common (1 in 100) blood clotting mutation. She was surprised when her provider responded by saying that, were she to become pregnant, she would need to be put on an injectable anti-coagulant throughout her pregnancy. With no family history of blood clotting disorders, nor personal history of recurrent miscarriage, this mutation would have gone untested, had it not been for the DTC results. However, when this patient did become pregnant and consulted with a specialist whose expertise was in blood clotting disorders in pregnant women, the anti-coagulant was indeed prescribed. Note that the guidelines have since changed, and the prophylactic treatment for clotting would not be prescribed today. It is also worth noting that in the United States, the Genetic Information Nondiscrimination Act (GINA) prevents employers and health insurers from discriminating against anyone based on their genetic information. It does not, however, cover long-term care, disability, or life insurance. Therefore when this woman applied for life insurance after her twins were born, the rate she was offered was more than twice what it would have been had she not known about the blood clotting mutation and been treated prophylactically for the increased risk it conferred.

A more positive example of where genetic testing is helping patients is a case presented at the American Neurological Association conference in 2014. A patient had a history of Alzheimer's disease on her mother's side of the family. She did not know if she was a carrier, nor did she want to know. But she wanted to ensure that she did not pass that mutation to her future children. Preimplantation genetic diagnosis (PGD) testing enabled her doctors to select embryos that did not have that Alzheimer's disease gene mutation. The patient herself was never tested, nor was she informed how many (if any) of the embryos contained the mutation.

Privacy

A 2013 Science paper from the Erlich lab at Massachusetts Institute of Technology (MIT) generated much controversy when the authors demonstrated the ability to re-identify a number of individuals using publicly-available genealogy databases and genetic data . The researchers used the short tandem repeat (STR) data from the Y chromosome, year of birth, and state of residence, combined with information from public genealogy websites to identify individuals. They did this by starting with publicly-available STRs and entering them on genealogy websites to identify matches. Note that their accuracy was not 100%. Though they were able to identify Craig Venter based on his genomic data, they failed to identify several other individuals, particularly those with more common last names. Overall, they reported a 12% success rate in recovering surnames of US males. They were also able to reconstruct Utah family pedigrees based on 1000 Genomes Project data and other publicly-available sources. Due to a number of cultural and historical factors, families in Utah tend to be large and genealogically well-documented, and an unusually high proportion of the Utah population has participated in scientific studies involving genomic data. It is worth noting that the researchers did not release any names that were not previously public, nor did they use the information for any nefarious purposes. Their primary interest was to demonstrate that such re-identification was possible. Interestingly, and somewhat surprisingly, the terms of use for the datasets did not prohibit re-identification.

Where are we going? The Road Ahead

Though we cannot know what the future holds, we can make some informed guesses based on events to date. The author believes that in the not-too-distant future, newborns will be sequenced at birth, just as we currently test for a more limited number of genetic issues. With the cost of sequencing a genome still at or above US $1000, such widespread sequencing is not yet realistic. But researchers are already performing pilot studies in this area, to better understand and anticipate the issues that are likely to arise. As an example, the MedSeq project at Harvard University is a study designed both to integrate WGS into clinical care and to assess the impact of doing so In addition, the Geisinger Health System has partnered with Regeneron on a project known as MyCode, which aims to sequence the exomes of 250,000 patients in the Geisinger system. In late 2015, the project began returning results to patients for 76 genes.

Even clearer is that tumor sequencing will be performed as part of standard of care for cancer. Currently sequencing is performed at certain tertiary and quaternary care facilities, particularly for metastatic tumors. As more is discovered about the various dysregulated pathways in cancer, and about the therapeutic implications for different genetic variations, the blunt mallet that is chemotherapy will be phased out in favor of far more precise and targeted therapies.

The microbiome has seen increasing attention in recent years, a trend that will certainly continue for the foreseeable future. It is not surprising that the make-up of the microbial communities that likely outnumber the cells of the human body can have significant impact on human health, particularly in metabolic and gastro-intestinal disease. The more surprising trend is the connection between the microbiome and other, more unexpected phenotypes, for example, anxiety, depression, and autism. This is likely to continue as this area of research continues to grow.

We will continue to see an increase in analyses of different "omic" types. Genomics has been by far the most popular area of focus to date. As technologies mature, we will continue to see biomarker discovery in proteomics, metabolomics, and other as-yet-unnamed "omic" modalities. We will also see increasingly integrated analysis, taking a systems approach to human biology where to date systems biology has been focused on model organisms, often single cellular ones, in which the system can be methodologically, and ethically, perturbed. Early examples of integrative, multi-modal analysis include the integration of microRNAs and transcription factors to determine regulatory networks underlying coronary artery disease, integrative analysis of genomics and transcriptomics to look at cardiovascular disease, and the use of metabolomics data with GWAS to elucidate molecular pathways.

The coming decade will see more biomarker-based research and insights into mental health disorders. To date, cancer and cardiology have received significant attention, to great advantage. But those disease areas are, by comparison, relatively easy to identify, to distinguish, and even to quantify. This is not the case for neurological and psychiatric disorders. Mental health is an area where diagnosis, and phenotyping more generally, is as much art as science. It is an area that poses enormous burdens on society, both financial and quality-of-life related, and is also ripe for a deeper, more physiologically-based understanding. Even if therapy is still a long way away, having some concrete, quantifiable biomarkers, by which we could classify conditions such as depression, bi-polar disorder, and manic-depressive tendencies, would be a great leap.

Finally, major changes will be required to effectively and efficiently train the workforce of tomorrow. These changes will not simply entail adding a few quantitative courses into medical and graduate level biomedical research training, though that too will be important.

As genomic sequencing becomes increasingly widespread, this number needs to increase both in the US and around the globe. Genetic counselors today may be compared to pathologists in the early days of the microscope, or radiologists in the early days of X-rays. The respective numbers will never need to be equal—except in the case of cancer, the genome need only be sequenced once in a person's lifetime. Moreover, pathologists and radiologists detect and describe what is. Even as we learn more about our genomes, they primarily tell us what is more or less likely to be. Still, many genetic findings are already actionable today, and this number will continue to increase.

There are two areas where we need to do better, but I am less optimistic that we will see real progress in the next decade or so. The first is in adoption of data standards. We need better resources for understanding, navigating, and using existing standards. We also need more impactful incentives for adoption, and disincentives for failure to do so. In the current landscape, standards are too difficult to identify and adopt, and the benefit of doing so tends to be realized by people other than those doing the hard work.

Lastly, we need to establish more inter-interdisciplinary coordination and collaboration. Perhaps meta-interdisciplinary is a better term. As biomedical informaticians, we are by definition interdisciplinary, including training and perspectives from medicine, biology, and computer science. But there are a large number of different communities around the world who are working on these problems, talking mostly among themselves. The various professional societies, even the interdisciplinary ones, have their respective meetings. There is some cross pollination, and some overlap in who attends the respective events but still not enough. There could be so much more, and we could make significant progress, reduce redundancy, and increase return on investment for research funding if these groups could be more consciously and proactively in sync.

Structural Bioinformatics

Structural bioinformatics is the branch of bioinformatics which specifically targeting at the analysis or prediction for three-dimensional structure of biological systems. Structural bioinformatics deal with biological systems at different scales, from molecules to cellular environment. Many types of cellular compartments, for example, the organization of the cell cytoplasm and the structural organization of the genome in its nuclear environment could be formalized by integrative structural bioinformatics analysis. Structural bioinformatics usually involves following steps:

- Collecting of biological data, either high throughput sequencing data or imaging data.

- Building computational model, can be structural simulation, optimization or alignment.

- Interpreting the model results from structural biology perspectives.

- Providing insights for next iteration of experimental design.

Recently developed technologies has enabled high throughput sequencing and imaging data collection. For example, Chromosome conformation capture techniques are used to analyze the spatial organization of chromatin in a cell based on high throughput sequencing data. By quantifying interactions between genome loci, we are able to infer the three-dimensional folding mechanism of the genome. For the imaging data, Cryo-electron tomography is an important tool to study structures of macromolecular complexes in close to native states based on 3D back-projection of a series of the projected 2D images of the structure.

Many statistical learning methods and computational algorithms could be applied towards building structural bioinformatics models. Molecular dynamics, for example, could be useful to simulate the distributions of biological molecules and the interactions among them. Clustering methods, could be applied to detect subtypes of conformations among population data. Convolution neural networks is a good fit for image processing. Graph-based learning methods is a suitable way to perform structural alignments. With statistical learning methods and computational algorithms, structural bioinformatics has the great potential to help save labor, cost and time. Moreover, it could help providing insights for next iteration of experimental designs so as to collect more biological data.

Applications of Bioinformatics

Varietal Information System

Bioinformatics has useful application in developing varietal information system. In connection with plant variety protection (PVP) Act, various terms such as extant variety, candidate variety, reference variety, example variety and farmer's variety are frequently used. Hence, knowledge of these terms is essential. These are defined below:

1. Extant Variety: All released, notified and unprotected varieties.

2. Candidate Variety: A variety to be registered under Plant Variety Protection Act is referred to as candidate variety.

3. Reference Variety: All released and notified extant varieties of common knowledge which are in seed production chain.

4. Example Variety: A variety that is used for comparison for a particular character is called example variety.

5. Farmers Variety: A variety that has been developed by a farmer and used for commercial cultivation for several years is called farmers variety.

The detailed information about various types of varieties can be developed using highly heritable characters.

Such information can be used in various ways as given below:

1. For varietal identification in DUS testing.

2. In grouping of varieties on the basis of various highly heritable characters.

3. In sorting out of cultivars for use in pre-breeding and traditional breeding.

The information can be stored in the computer memory and be retrieved as and when required.

Plant Genetic Resources Data Base

The genetic material of plants which is of value for present and future generations of people is referred to as plant genetic resources. It is also known as gene pool, genetic stock and germplasm. The germplasm is evaluated for several characters such as highly heritable morphological, yield contributing characters, quality characters, resistance to biotic and abiotic stresses and characters of agronomic value.

International Plant Genetic Resources Institute (IPGRI), Rome Italy has developed descriptors and descriptor states for various crop plants. Such descriptors help in uniform recording of observations on germplasm of crop plants all over the world. Thus huge data is collected on crop genetic resources for several years. Bioinformatics plays an important role in systematic management of this huge data.

Bioinformatics is useful in handling such data in several ways as follows:

1. It maintains the data of several locations and several years in systematic way.

2. It permits addition, deletion and updating of information.

3. It helps in storage and retrieval of huge data.

4. It also helps in classification of PGR data based on various criteria.

5. It helps in retrieval of data belonging specific group such as early maturity, late maturity, dwarf types, tall types, resistant to biotic stresses, resistance to abiotic stresses, superior quality, marker genes, etc.

All such data can be easily managed by computer aided programs and can be manipulated to get meaningful results.

Biometrical Analysis

In crop improvement, various biometrical analyses are performed.

Important biometrical analyses that are performed in plant breeding and genetics are given below:

1. Simple measures of variability such as mean, standard deviation, standard error, coefficient of variation, etc.

2. Correlations: It includes genotypic, phenotypic and environmental correlations. It also includes simple, partial and multiple correlations.

3. Path Coefficients: It includes analysis of genotypic, phenotypic and environmental paths.

4. Discriminant function analysis.

5. Metroglyph analysis and D2 statistics.

6. Stability analysis.

7. Diallel, partial diallel, line x tester triallel, quadriallel, biparental and triple test cross analysis.

8. Generation mean analysis, etc.

All these analyses can be easily performed through computer aided programs.

Storage and Retrieval of Data

In crop improvement, huge data is collected on the following aspects:

1. Segregating populations:

Single plant selections are made in segregating populations and data are recorded on various characters such as yield components, quality characters, resistance to biotic and abiotic stresses, etc.

2. Multi-location Experiments:

Such experiments are conducted mainly for identification and release of new varieties and hybrids and also for assessment of varietal stability.

3. Multi-seasonal Experiments:

Such experiments are conducted for several years (3-5 years) for identification of new varieties and hybrids. The above data remain in active use generally for two decades. Handling of such a huge data is a difficult task.

However, such data can be easily stored in various storage devices such as hard disks, compact disks, pen drive, data cards, etc. Storage of data in computers require less space and is very safe as compared to storage of data in paper registers and files.

Studies on Plant Modelling

Computers are useful tools for undertaking studies on modelling of plants. First the theoretical model can be prepared with the help of computer keeping in view various plant characters. Then such model plants can be developed through hybridization and directional selection.

This type of studies is useful in developing crop ideotypes or ideal plant types in different field crops. First the conceptual model is prepared and then efforts are made to achieve such model by combining desirable genes from different sources into a single genotype through appropriate breeding procedures. Such studies have been made in field pea.

Pedigree Analysis

Computer aided studies are useful in pedigree analysis of various cultivars and hybrids. Information about the parentage of cultivars and hybrids is entered into the computer memory which can be retrieved any time. The list of parents which are common in the pedigree of various cultivars and hybrids can be sorted out easily.

It helps in the pedigree analysis which in turn can be used in planning plant breeding programs especially in the selection of parents for use in hybridization programs. The study of proteomics also helps in pedigree analysis.

Preparation of Reports

After biometrical analysis of data, results are interpreted and various types of reports or documents are prepared.

In crop improvement following types of reports is prepared:

1. Research Project Report: The annual progress report of each project is prepared and salient findings are documented.

2. Monthly, quarterly, half yearly and annual progress reports of all the research projects are also prepared.

3. Sometimes, bulletin and booklets are prepared to document specific information for adoption and benefit of farmers.

4. Research papers and popular articles are prepared based on research findings.

5. Germplasm catalogues are prepared for various characters.

Such reports can be easily be prepared with the help of computers using MS Word program. This information can be stored in computer memory and reused as and when required. The editing and updating of reports can be done any time without much extra efforts.

Updating of Information

In plant breeding and genetics, results of multi-seasonal and long term experiments require continuous updating. Computers have made this task very simple. The information related to any experiment which is already stored in the computer memory, can be updated any time by editing the concerned file. Any portion of information can be deleted or revised easily.

Diagrammatic Representation

Inclusion of diagrams makes the reports, research papers, articles, bulletins, etc. more attractive, informative and easily understandable.

The following types of diagrams are made in plant breeding:

1. Line diagrams, bar diagrams, histograms and pie diagrams.

2. Cluster diagram: It is prepared when data is subjected to D2 analysis.

3. Path diagram: It is prepared when data is subjected to path coefficient analysis.

4. Vr-Wr Graph: It is prepared when data is subjected to Hayman's graphical approach of diallel cross analysis.

5. Metroglyph Chart: It is prepared when data is subjected to Metroglyph analysis.

All these diagrams can be easily prepared with the help of computer using specific program.

Bioinformatics in Plant Breeding

Plants are the basis of life on earth. They produce the life-supporting oxygen we breathe, they are essential for our nutrition and health and they provide the environment for the vast biodiversity on earth. For centuries, humans have selected plant varieties that best fit their purposes and developed crop plants that have many advantages compared to natural (wild) plants in quality, quantity and farming practises. However, multifactorial traits involved in resistance and quality have proven to be extremely difficult to improve, certainly in combination. The revolution in life sciences signalled by genomics dramatically changes the scale and scope of our experimental enquiry and application in plant breeding. The scale and high resolution power of genomics enables to achieve a broad as well as detailed genetic understanding of plant performance at multiple levels of aggregation. The complex biological processes that make up the mechanisms of pathogen resistance and provide quality to our crops are now open for a systematic functional analysis. These analysis are made with specific software on the high amounts of data generated in databases and is the field of plant bioinformatics.

Genome initiatives are under way for more than 60 different plant species. From the point of view of economics, the most important of these are those of the major feed crops – the grasses maize, rice, wheat, sorghum and barley; and the forage legumes soybean and alfalfa. Several of these genomes are so large (as result of autopolyploidization and the dramatic expansion of repetitive DNA) that whole genome sequencing is impractical, and efforts have instead been focused on comparative genome methods. Both rice and maize, however, have relatively small genomes and are such key elements of the agricultural economies of the developed world that complete genome sequences have been prioritazed.

Classification of databases in the 2004 edition of the Molecular Biology Database Collection.

Category	No of databases
Genomic	164
Protein sequences	87

Human/vertebrate genomes	77
Human genes and diseases	77
Structures	64
Nucleotide sequences	59
Microarray/gene expression	39
Metabolic and signaling pathways	33
RNA sequences	32
Proteomics	6
Other	16

The role of model organism. Over the last century, research on a small number of organisms has played a pivotal role in advancing our understanding of numerous biological processes. This is because many aspects of biology are similar in most or all organisms, but it is frequently much easier to study a particular aspect in one organism than in others. These much-studied organisms are commonly referred to as model organisms, because each has one or more characteristics that make it suitable for laboratory study. The most popular model organisms have strong advantages for experimental research, such as rapid development with short life cycles, small adult size, ready availability, and tractability, and become even more useful when many other scientists work on them. A large amount of information can then be derived from these organisms, providing valuable data for the analysis of normal human or crop development; gene regulation, genetic diseases, and evolutionary processes.

Comparison of genome sequences of rice and Arabidopsis suggests that extensive but complex patterns of synteny will be a useful feature of plant genomics. Medicago (alfalfa) is a true diploid that, along with its crucial role in the fixing soil nitrogen, constitutes a major part of forage diets. Other grasses and legumes are the subjects of extensive EST sequencing and high resolution genetic map construction, in some cases involving radiation hybrid mapping, in hopes of taking advantage of the expected pervasive synteny within these families. Web sites established by individual research groups integrate research efforts from around the globe. In the 1980s, there was a growing awareness that significant investments in studies of many different plants, such as corn, oilseed rape, and soybean, were diluting ef forts to fully understand the basic properties of all plants. Scientists began to realise that the goal of completely understanding plant physiology and development is so ambitious that it can best be accomplished by turning to a model plant species that many scientists then study. Fortunately, because all flowering plants are closely related, the complete sequencing of all the genes of a single, representative, plant species will yield much knowledge about all higher plants. Similarly, discovery of the functions of the proteins produced by a model species will offer much information about the roles of proteins in all higher plants.

Arabidopsis thaliana has become universally recognised as a model plant for study. It is a small flowering plant that belongs to the Brassica family, which includes species such as broccoli, cauliflower, cabbage, and radish. Although it is a non-commercial plant, it is favoured among basic scientists because it develops, reproduces, and responds to stress and disease in much the same way as many crop plants. Scientists expect that systematic studies of Arabidopsis will offer important advantages for basic research in genetics and molecular biology and will illuminate numerous features of plant biology, including those of significant value to agriculture, energy, environment, and human health. Because of several reasons Arabidopsis has become the organism of choice for basic studies of the molecular genetics of flowering plants.

Arabidopsis thaliana has a small genome (125 Mb total), which already has been sequenced in the year 2000, and it lacks the repeated, less-informative DNA sequences that complicate genome analysis. It has extensive genetic and physical maps of all 5 chromosomes (MapViewer); a rapid life cycle (about 6 weeks from germination to mature seed); prolific seed production and easy cultivation in restricted space; efficient transformation merits utilising Agrobacterium tumefaciens; a large number of mutant lines and genomic resources (Stock Centers) and multinational research community of academic, government and industry laboratories.

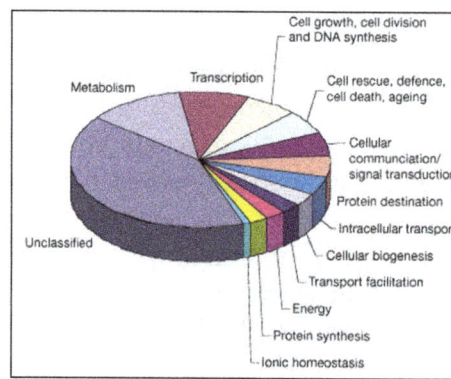

During the Arabidopsis evolution the whole genome has duplicated once, followed by subsequent gene loss and extensive local gene duplications. The genome contains 25.498 genes encoding proteins from 11.000 families. The genome is compared to previous sequenced genomes and ESTs, and as much functions of genes are predicted. This functional analysis of the Arabidopsis genome showed the following proportion of predicted function. As with other model organisms there is much more to the Arabidopsis genome project than the complete genome sequence. The Web site for the Arabidopsis Information Resource, TAIR, allows researchers to integrate the genome sequence with an extensive EST data base and with the genetic and physical maps; it provides links to functional and molecular genetic data and the literature for specific genes; and it shows an ever expanding list of mutant stocks. An alternative resource for Arbabidopsis and many other plants UK CropNet, uses common AceDB or WebAce platforms to coordinate genetic and molecular data.

As plant model organisms also could be considered:

- Medicago truncatula (commonly known as "barrel medic" because of the shape of its seedpods) is a forage legume commonly grown in Australia.

- Tomato: The overall goal of the Tomato Genomics Project includes the development of an integrated set of experimental tools for use in tomato functional genomics. The resources developed will be used to further expand our understanding of the molecular genetic events underlying fruit development and responses to pathogen infection, and will be made available to the research community for analysis of diverse plant biological phenomena.

- Rice: Some desired features of future improved rice varieties are superior grain quality, higher yield potential, enhanced resistance to insect pests and diseases, and greater tolerance to stresses such as drought, cold, and nutrient deficiencies.

- Maize: Maize products produce about $30 billion every year, and is used for food, rubber, plastic, fuel, and clothing. The maize genome is about 20 times larger than the one from Arabidopsis. This means that it is as big as the human genome. However, it's organisation is more complex to sequence than that of all organisms that are sequenced today. The genes are situated in clusters through the genome, with high amounts of repetitive sequences in between. The genes containing regions make up to 15% of the total genome. Other significant characteristics of the maize genome are that it contains multiple copies of most genes and the existance of jumping genes or transposons that make up a large portion of the genome.

- Wheat: Recent advances in plant genetics and genomics offer unprecedented opportunities for discovering the function of genes and potential for their manipulation for crop improvement. Because of the large size of the wheat genome, it is unlikely that the actual base pair sequences of the DNA molecules will be learned completely in the near future. This project takes an alternative strategy to realise the benefits of new techniques for discovering genes and learning their function (functional genomics). Following the identification of 10,000 wheat ESTs, they will be mapped to their physical location on the chromosomes of wheat. This process utilizes a unique feature of the wheat chromosomes, their ability to tolerate deletions of portions of the chromosomes and still produce a viable plant. The mapping logic is direct: if an EST is present in a plant with complete chromosomes, but absent in a plant missing a known part of a single chromosome, then it can be inferred that the DNA sequence that corresponds to that EST is located in that segment of the chromosome. By the end of the mapping component of this project, a most valuable tool will have been produced: 10,000 unique DNA sequences, likely corresponding to genes, whose physical location in the chromosomes of wheat are known. This sets the stage for the next phase of the project, the analysis of this array of mapped ESTs to determine function.

- Other flowering plants: Over 90 different angiosperm genome projects around the world are listed on the United States Department of Agriculture Web site The list includes African projects on beans, corn and fungal pathogens; Australian projects on cotton, wheat, pine, sugarcane, and nine others; at least 24 European projects that include vegetables such as cabbage, cucumber, and pea, and fruits such as apple, peach and plum; and over 50 North American projects as diverse as turf grass, chrysanthemum, almond, papaya grape and poplar. The common denominator among all of these projects is the assembly of genetic maps (and in some cases physical maps) and the placement of a common set of plant genes on them. For some species large EST sequencing projects are also in place with the twin objectives of enabling comparative genomic analysis (particularly in the regions of synteny) and QTL mapping.

Managing and Distributing Plant Genome Data

As with many areas of science and technology genome science has benefited greatly from advances in computing capabilities and bioinformatics. Improved computational speed has been important but a strong argument can be made that the growth of the Internet has been even more crucial for genome scientists. In conjunction with the maturation of modern database technology the World Wide Web has become the natural medium for managing and distributing genomic resources.

The emergence of the Internet allowed the creation of centralized data warehouses. Just as important, it led to the creation shared public resources for searching and analyzing the contents of genomic databases. Full-featured Web sites such as those at NCBI and EMBL provide immediate access to enormous amounts of data and analysis tools, free of charge, from anywhere of the globe. This is dramatic change from the situation about a decade go, when GenBank database was distributed by paid subscription in a small notebook, full of 5.25" floppy disks.

Networking advances have also been important for within laboratory data management and with little or no human intervention. Centralized laboratory information management systems or LIMS, then allow users at multiple workstations or even multiple geographic locations to browse, edit, analyze and annotate the data.

The core item of genomic data is a database system. Most databases can be classified as either relational databases (RDB) or object-oriented data bases (OODB).

There are three primary sequence databases: GenBank (NCBI), the Nucleotide Sequence Database (EMBL) and the DNA Databank of Japan (DDBJ). These are repositories for raw sequence data, but each entry is extensively annotated and has features table to highlight the important properties of each sequence. The three databases exchange data on a daily basis. Similarly, SWISS-PROT and TrEMBL are the major primary databases for the storage of protein sequences. There are also secondary databases of protein families and sequence patterns such as PROSITE, PRINTS and BLOCKS. These are called secondary databases because the sequences they contain are not raw data, but they have been derived from the data in the primary databases.

The early bioinformatics databases emphasized primary data capture. GenBank, established in the late 1980's began with staff at Los Alamos National Laboratoty (and later the National Center for Biotechnology Information) manually keying DNA sequences from published papers into the computer. It neither added nor removed any information from these sequences, nor did it perform any integration of multiple overlapping sequences. Other databases founded around this time also focused on a single data type: SwissProt for protein sequences and PDB for X-ray crystallographic structures. From the mid-1990s to the early part of this decade the emphasis shifted from data capture to data aggregation and integration. This was largely due to the limitations of primary data archives: what if one wanted to correlate DNA sequencing information with data from biochemical or genetic studies? Model Organism Databases (MODs), integrated repositories of all the electronic information resources pertaining to a particular experimental plant or animal species, became the darlings of the bioinformatics world.

Now the MOD paradigm is itself giving way to new, higher level concepts, such as clade-specific and pathway databases. Integrating multiple types of biological data across several species, these resources enable researchers to make discoveries that wouldn't be possible by examining a single species alone. It could be predicted that in another decade the idea of biological database devoted to one species will seem as quaint as the idea of a database devoted to a single type of laboratory data seems today.

Though MOD are still going strong, their preeminence is now being challenged by multispecies, comparative-genomics databases, sometimes called clade-specific databases. These systems integrate information on multiple organisms and use comparative analysis to discover patterns in genome that might otherwise be missed. Well known clade specific databases include: EnsEMBL at the European Bioinformatics Institute (EBI; Entrez at the NCBI and the Genome Browser at the University of California, Santa Cruz all of which relate information on the human genome to data gathered over plants, vertebrates, invertebrates and prokaryotes. Some years ago the Cold Spring Harbor Laboratory, New York established Gramene a comparative genomics resource for crop grasses. This database integrates genome sequences, genetic maps, mutation and trait data across rice, maize wheat and a large number of other cereals. Gramene gives researchers the benefit of genome sequencing even before their favorite organism actually has been sequenced.

The maize genome, for instance, is about the same length as the human genome, and won't be fully sequenced for another several years, but rice, with a compact genome one-tenth the size of human's, already is. Because the two grains are closely related evolutionarily, we have been able to create maps that relate maize's genetic map to the rice genome sequence. This allows to researchers to follow a genetically mapped trait in maize, such as tolerance to high salt levels in the soil, and move into the corresponding region in the rice genome, thereby identifying candidate genes for salt tolerance. Similar techniques helped cattle researchers identify in 1997 a gene responsible for muscle growth based on the existence of a genetic mutation in a corresponding region of the mouse genome.

The next very important class of databases in near future could be considered as pathway databases. Traditional databases are linear catalogues of sequences, genes, proteins, genomes, and genome-to-genome alignments. Such databases have one or a small number of central data objects, such as gene record, and all the other information hangs off that object. A typical research project describes the model of the series of experiments. The model usually describes the series of molecular events (the pathway) that is responsible for whatever phenomenon is studied, whether it be embryonic development, neuronal signaling in the brain, or the transformation of healthy cells into cancerous ones. These pathways are the ultimate output of biological research. The initial project in pathway databases is the Reactome Current entries describe energy metabolism, DNA replication, RNA transformation and splicing, protein translation and cell cycle regulation. Each pathway is linked to the literature references that provide experimental support for it, and to the database records for genes, sequences and proteins that participate in the pathways.

A new standard in transferring metabolomics data has been developed since 2002 by several BioPAX project, and financed by Department of Energy of United States. The goal of the BioPAX group is to develop a common exchange format for biological pathways data.

Sequence Alignment Methods and Applications

Comparing genome sequences. The development of technologies for the largescale quantification and identification of biological molecules combined with advances in computing technologies and the internet has served to facilitate the delivery of large volumes of biological data to the scientists' desktop. By the time the human genome sequence was published in 2001, the rate of DNA sequencing had increased 2,000-fold since the inception of the technology in 1986. The increased productivity was gained through automation, miniaturization, and integration of technologies; applying this approach to the analyses of other biological molecules including mRNA, proteins, and metabolites has resulted in a massive increase in the generation of biological data. This data has been made easily accessible, in part due to publications such as the Molecular Biology Database Collection, an annual listing of the best databases publicly available to the biological community. Analysis of the collection reveals the steady growth in the quality and size of the databases with the 2004 edition containing 548 databases classified into 11 categories.

DNA sequencing is performed using and automated version of the chain termination reaction, in which limiting amounts of dideoxyribonucleotides generate sets of DNA fragments with specific terminal bases. Four reactions are set up, one for each of the four bases in DNA, each incorporating different fluorescent label. The DNA fragments are separated by the PAGE and the sequence is read by a scanner as each fragment moves to the bottom of the gel.

DNA sequences come in three major forms. Genomic DNA comes directly from the genome and includes extragenic material, as well as genes. In eukaryots, genomic DNA contains introns. cDNA is reverse transcribed from mRNA and corresponds only to expressed parts of the genome. It does not contain introns. Finally, recombinant DNA comes from the laboratory and comprises artificial DNA molecules such as cloning vectors.

Major aim of most genome projects is to determine the DNA sequence either of the genome or of a larger number of transcripts. This endeavor both leads to the identification of all or most genes and to the characterization of various structural features of the genome. Very often the major essence of the bioinformatics strategies for sequence alignment is the comparison of cDNA/EST and genomic sequences and annotation. The veracity of any whole genome sequence must be assessed at three levels: its completeness, the accuracy of the base sequence and the validity of its assembly.

In addition to whole genome sequencing, plant sequence data have been accumulating from three major sources: sample sequencing of bacterial artifcial chromosomes (BACs), genome survey sequencing (GSS) and sequencing of expressed sequence tags (ESTs).

1. Quence tags (ESTs): Sequence alignment is the arrangement of two or more amino acid or nucleotide sequences from an organism or organisms in such a way as to align areas of the sequences sharing common properties. The degree of relatedness or homology between the sequences is predicted computationally or statistically based on weights assigned to the elements aligned between the sequences. This in turn can serve as a potential indicator of the genetic relatedness between the organisms.

2. Alignment is a computational problem: There is a certain degree of convinction, that two similar sequences can be lined up in such a way that identical bases (or amino acids) are all matched. However from a computers point of view the alignment process is far from trivial. If gaps are allowed there are a tremendous number of of different alignments possible for any two sequences.

Dynamic programming algorithms can calculate the best alignment of two sequences. Well known variants for pairwise alignment are the Smith-Waterman algorithm for local alignment and the

Needelman-Wunsch (19) algorithm for global alignment. Local alignments are useful when sequences are not related over their full lengths, for example proteins sharing only certain domains, or DNA sequences related only in exons.

Multiple sequence alignment. Multiple alignment illustrates relationships between two or more sequences. When the sequences involved are diverse, the conserved residues are often key residues associated with maintenance of structural stability or biological function. Multiple alignments can reveal many clues about protein structure and function. The most commonly used alignment software is the ClustalW package.

Alignment scores and gap penalties. A simple alignment score measures the number of proportion of identically matching residues. Gap penalties are subtracted from such scores to ensure that alignment algorithms produce biologically sensible alignments without too many gaps. Gap penalties may be constant (independent of the length of the gap), proportional (proportional to the length of the gap) or affine (containing gap opening and gap extension contributions). Gap penalties can be varied according to the desired application. Sequence similarity can be quantified using score from the alignment algorithm, percentage sequence identities or more complex measures.

Sequence Similarity Searching Algorithms

Dynamic programming algorithms are guaranteed to find the best alignment of two sequences for given substitution matrices and gap penalties. This is impressive, but the process is often quite slow, perhaps taking hours for a search of a large database. For these reason alternative methods have been developed. Perhaps the most used of these are FASTA Both tools BLAST and FASTA provide very fast searches of sequence databases. Unlike dynamic programming, they do not guarantee to find the best possible alignment to each database sequence, but in practice the effect of performance is usually minimal. Each operates by first locating short stretches of identically or near identically matching letters (words) that are eventually extended into longer alignments. Best BLAST server runs the NCBI and can be used to search many general-purpose sequence databases. A similar FASTA implementation is available at the EBI.

Smith-Waterman is an algorithm for local sequence alignment, using as input two sequences. The difference between NCBI BLAST (also considered as local alignment algorithm) and Smith-Waterman is that a) BLAST is searching query sequence over a database of sequences; and b) BLAST is calculating statistically most probable alignment match, since SmithWaterman is calculating the exact match.

Genome Comparison Tools. MegaBlast is NCBI BLAST based algorithm for large sequence similarity search. MegaBlast implements a greedy algorithm for the DNA sequence gapped alignment search. MegaBlast is used to compare the raw genomic sequences to a database of contaminant sequences (including the UniVec database of vector sequences, the Escherichia coli genome, bacterial insertion sequences, and bacteriophage databases). Any foreign segments are removed from draft-quality sequence or masked in finished sequence to prevent them from participating in alignments.

Jim Kent's BLAT (BLAST-Like Alignment Tool) is a tool which performs rapid mRNA/DNA and cross-species protein alignments. BLAT is more accurate, 500 times faster than popular existing algorithms for mRNA/DNA alignments, and 50 times faster for protein alignments at sensitivity settings typically used when comparing vertebrate sequences.

Genome based multiple alignment using BlastZ. BLASTZ is a multiple sequence alignment program basically used for the whole-genome human-mouse alignments. Blastz output can be viewed with the LAJ interactive alignment viewer, converted to traditional text alignments. LAJ is a tool for viewing and manipulating the output from pairwise alignment programs such as BLASTZ. It can display interactive dotplot, pip, and text representations of the alignments, a diagram showing the locations of exons and repeats, and annotation links to other web sites containing additional information about particular regions.

EST sequencing ESTs are partial gene sequences which have been generated or are in the process of being produced in several laboratories using different species and cultivars as well as varied tissues and developmental stages. This represents an important step towards the identification of all expressed genes for instance in grapevine, and some members of the Rosaceae family: a large part of the Malus (apple) genome, raspberries/blackberries Rubus, stone fruits (Prunus), strawberry (Fragaria) peach, almond.

ESTs are now widely used throughout the genomics and molecular biology communities for gene discovery, mapping, polymorphism analysis, expression studies, and gene prediction Expressed sequence tags (ESTs) have applications in the discovery of new genes, mapping of the genome, and identification of coding regions in genomic sequences. An EST database consists of ESTs drawn from multiple cDNAs, and there could be potentially many ESTs drawn from each cDNA. Given such a database, the EST clustering problem is defined as follows: The ESTs should be partitioned into clusters such that ESTs from each gene are put together in a distinct cluster. A further complication arises due to the fact that DNA is a double stranded molecule and a gene could be part of either strand.

dbEST is a division of GenBank that contains sequence data and other information on "single-pass" cDNA sequences, or Expressed Sequence Tags, from a number of organisms. The Institute for Genomic Research (TIGR) is defining also TC as Tentative Consensi (assemblies from ESTs) and ET as Expressed Transcripts (both non-human) when building TIGR Gene Indices (TGI).

Molecular Information and Plant Breeding – A Bioinformatics Approach

Molecular plant breeding. As the resolution of genetic maps in the major crops increases, and as the molecular basis for:

- Number of ESTs by fruit collected in dbEST (release 040805).

- dbEST release 040805.

- Number of public entries: 26,605,325.

Malus x domestica	183916
Vitis vinifera	147300
Prunus armeniaca	15181
Citrus x paradisi x Poncirus trifoliata	8002
Vitis hybrid cultivar	6533
Fragaria x ananassa	5322

Prunus dulcis	3864
Citrus reticulata	3735
Citrus unshiu	2561
Ananas comosus	1547
Fragaria vesca	1306
Citrullus lanatus	693
Citrus clementina x Citrus reticulata	74
Vitis cinerea x Vitis rupestris	61
Cucumis melo	60

Specific traits or physiological responses becomes better elucidated, it will be increasingly possible to associate candidate genes, discovered in model species, with corresponding loci in crop plants. Appropriate relational databases will make it possible to freely associate across genomes with respect to gene sequence, putative function, or genetic map position. Once such tools have been implemented, the distinction between breeding and molecular genetics will fade away. Breeders will routinely use computer models to formulate predictive hypotheses to create phenotypes of interest from complex allele combinations, and then construct those combinations by scoring large populations for very large numbers of genetic markers.

The vast resource comprising breeding knowledge gathered over the last several decades will become directly linked to basic plant biology, and enhance the ability to elucidate gene function in model organisms. For instance, traits that are poorly defined at the biochemical level but well es tablished as a visible phenotype can be associated by high resolution mapping with candidate genes. Orthologous genes in a model species, such as Arabidopsis or rice, may not have a known association with a quantitative trait like that seen in the crop, but might have been implicated in a particular pathway or signaling chain by genetic or biochemical experiments. This kind of cross-genome referencing will lead to a convergence of economically relevant breeding information with basic molecular genetic information. The specific phenotypes of commercial interest that are expected to be dramatically improved by these advances include both the improvement of factors that traditionally limit agronomic performance (input traits) and the alteration of the amount and kinds of materials that crops produce (output traits). Examples include:

- Abiotic stress tolerance (cold, drought, salt).

- Biotic stress tolerance (fungal, bacterial, viral, chewing and sucking insect attack (feeding)).

- Nutrient use efficiency - manipulation of plant architecture and development (size, organ shape, number, and position, timing of development, senescence).

- Metabolite partitioning (redirecting of carbon flow among existing pathways, or shunting into new pathways).

Rational plant improvement. The implications of genomics with respect to food, feed and fibre production can be envisioned on many fronts. At the most fundamental level, the advances in genomics will greatly accelerate the acquisition of knowledge and that, in turn, will directly impact many aspects of the processes associated with plant improvement. Knowledge of the function of all plant genes, in conjunction with the further development of tools for modifying and interrogating genomes, will lead to the development of a genuine genetic engineering paradigm in which rational changes can be designed and modeled from first principles.

Genotype Building Experiments

Biodiversity determined by the plant genome analysis. In the recent years an increasing amount of information for the DNA polymorphism and sequencing was accumulated in different plant varieties and cultivars. Most of this information was used for the purpose of recognition of different cultivars as well as for their comparison – distances and similarities (22). These distances are measured by the polymorphism on a part of the chromosome with unknown function. This type of polymorphism is widely used in the genomic studies across the species. The data for the polymorphism are analyzed for a possible link with a quantitative trait of interest of the individual phenotypes. Once such a link is detected it is called indirect marker.

Indirect markers are closely linked, sometimes they may overlap, with a locus which determine this quantitative trait – QTL. QTLs are defined as genes or regions of chromosomes which affect a trait. QTLs by themselves are difficult to be recognized. In both cases this information, or as it is called – markers, can be used in further selection purposes. This selection process is named as MAS.

QTLs and mapping. The major problem is to define which populations are suitable for QTL-analyses – unstructured and F2 crosses and in plant - large scale populations in order to screen for possible QTLs. As selection is based most on markers, higher density of mapping is important. The interval between marker and QTL of about 5 centiMorgans (cM) seemed sufficient for effective selection. The simulation studies however showed that selection accuracy dropped down to 81% and 74% with 2 cM and 4 cM distance compared to 1cM.

How QTL information could be of use?

- It is assumed that some but not all loci are identified, so selection should be based on the combination of phenotypic and molecular information;

- In the process of selection the link of markers and traits could decrease so this link should be observed throughout the generations;

- In the process of selection QTLs prove simultaneous existence of the desired genes in a line;

- In crossbred programs QTLs could predict the productivity of untested crosses, including their non-additive effect on the information of the parent lines and limited number of crosses;

- Future prospective – with accumulation of molecular data genotype building programs will be developed which will set homozygous desirable markers;

- In intrigression programs for combining the desirable traits from two lines in one;

- Finally – the real world of agriculture is on the stage of accumulation of molecular data.

Analytical approaches: One of the statistical tools for performing the QTL analyses such is the meta-analysis, which synthesize dense QTL information and refines the QTL position. A program of this class is the French BioMercator. An environment with complex research opportunities is also PlaNet – the European plant genome database network.

Further development and efficiency of QTLs and MAS. Further development and detailed discussion on QTLs includes the statistical aspects of MAS, setting up the threshold of significance of marker effects, overestimation or bias in estimation of QTL effects, optimization of selection programs for several generations with simultaneous utilization of MAS and phenotypic data. A specific feature is that detection should be made on plant specific parts – leaves, roots, fruits etc., as it was proved for the grapes.

Experimental results not always confirm the efficiency of MAS over the genotype building. The main reason is in unsufficient precision of the initial assessment of a QTL, its place and effect. Some QTLs also could be lost in the genotype building process. For complex productivity traits the epistatic lost would be a reason for changes in the magnitude of QTL effect in the parent and progeny generation. Then it is recommended that selection is based on the allelic combinations rather on the separate QTLs. It is in line with the numerous GxE interactions and with the selection within the environment of interest in the case of disease/drought resistance.

Consequently, efficiency of MAS will depend on the complexity of species/trait genetic architecture, on the development of the trait in the environment and on their interaction. For complex traits the evaluation of QTLs should be in different environments. Phenotypic evaluation/check throughout the consecutive generations is also necessary. For instance: drought resistance seemed to be more complex trait vs. disease resistance.

From the economics point of view the use of markers will cost collection of DNA, genotyping, analyses, detection of QTLs etc. This high price is paid for the genotype building (there is no other way of doing that), for traits that are expensive for evaluation – disease resistance, or traits with low heritability.

Role and Application of Bioinformatics in Plant Disease Management

Genomic studies focused on whole genome analysis, have opened up a new era for biology in general and for agriculture in particular. Along with the use of genetic plant models and the progress in sequencing agriculturally important organisms, the combination of bioinformatics and functional genomics globally enhance agricultural genomics. These studies are likely to pave the way towards better understanding of plant-pathogen biological network, and eventually to lead to break thoughts in promotion of plant resistance to pests. Bioinformatics plays a great role in plant disease management. Some of the areas in which bioinformatics is involved in plant diseases management are indicated.

Study of host- pathogen interaction The interaction between plants and their pathogens is complex. The study of these interactions has a long and rich history in science, with plant pathologists tackling these complex systems first with classical tools, such as physiology, histology, microbiology, plant breeding and genetics, and more recently with advanced biochemistry and molecular biology approaches. Plant pathogens have evolved a broad set of proteins that enable a stealthy entry into the plant cell and facilitate the evasion of host defenses. Among other defenses, plants have evolved a series of proteins that monitor their cells for signs of infection. Downstream of these monitors is a signaling and response system triggered upon infection.

The molecular basis of the host-pathogen interaction is now much better understood, as a result of the development of genomic data and tools. For example, the complete genomic sequence is available for a model plant, Arabidopsis, and for one of its bacterial pathogens, Pseudomonas syringaepv. tomato DC3000. Detailed molecular analyses of these two organisms have revealed much about plant defenses. Modern genomics tools, including applications of bioinformatics and functional genomics, allow scientists to interpret DNA sequence data and test hypotheses on a broader scale than previously possibl. In the last 5 to 10 years, many of the critical host proteins that detect the presence of pathogens have been characterized. Numerous components of the plant signaling system have also been identified that function downstream of the detection molecules. In parallel, the pathogen proteins that are used to suppress host defenses and drive the infection process (so-called effector proteins) have also been identified, using molecular biological technologies and genetics.

Study of Disease Genetics

Advances in molecular biology, plant pathology, and biotechnology have made the development of different techniques are designed to detect plant diseases early, either by identifying the presence of the pathogen in the plant (by testing for the presence of pathogen DNA) or the molecules (proteins) produced by either the pathogen or the plant during infection. These techniques require minimal processing time and are more accurate in identifying pathogens. In the past, disease genetics has focused on monogenic diseases in which the expression of a particular variant of a single gene will, in the vast majority of cases, lead to disease. There are innumerable monogenic diseases, each of which affects only a small number of individuals. In contrast, disease genetics research is now focused on identification of genes associated with common diseases. These common diseases are multi-factorial (i.e. dependent on complex interactions between numerous environmental factors and a number of alternative forms of genes called disease susceptibility genes) and polygenic (involving more than one gene in their multi-factorial pathogenesis. The overall goal of disease genetics is to identify how genetic variation can influence disease susceptibility and to improve understanding of the molecular processes resulting in clinically overt disease. New treatments can then be designed to target these molecular processes to prevent and treat the disease.

Typically, new disease susceptibility genes have been identified using a combination of linkage and association studies. The linkage studies involve collection of DNA samples and extensive clinical phenotypic data from multiple members of affected families. Markers are typed throughout the genome, and using linkage analysis algorithms, chromosomal regions harboring disease genes are identified. The regions are identified using highly informative markers on the basis of their chromosomal location by taking advantage of the meiotic process of recombination as apparent in families segregating for the disease. Markers closest to the disease gene show the strongest correlation with disease patterns in families. These linkage studies allow identification of a region on a chromosome and large portions of the DNA that may be linked to a specific disease.

Identification of Pathogencity Factors of a Pathogen

Pathogensity is the capacity of a pathogen to overcome genetically determined host resistance. The identification of genes involved in host–pathogen interactions is important for the elucidation of mechanisms of disease resistance and host susceptibility. A traditional way to classify the origin of genes sampled from a pool of mixed cDNA is through sequence similarity to known genes from either the pathogen or host organism or other closely related species. This approach does not work when the identified sequence has no close homologues in the sequence databases.

It was reported that bioinformatics play a role in identification of the HrpL region and type III Secretion System effector proteins of Pseudomonas syringaepv. phaseolicola1448A. The HrpL alternative sigma factor activates the expression of multiple genes that are essential to the plant pathogencity of Pseudomonas syringae. The most important of these are genes encoding the type III secretion system (T3SS) and effector proteins that are injected by the T3SS into host cells. The T3SS is encoded by hypersensitive response and pathogencity (hrp) and hrpconserved (hrc) genes. Effector proteins are encoded by avirulence (avr) and Hrp outer protein (hop) genes, whose different names reflect the phenotype used to discover them. Mutations were constructed in the DC3000 homologs and found to reduce bacterial growth in host These results establish the utility of the bioinformatic or candidate gene approach to identifying effectors and other genes relevant to pathogenesis in pathogen genomes.Similarly, it was reported that identification and characterization of potato protease inhibitors able to inhibit pathogencity and growth of Botrytis cinerea.

Developing Disease Resistance Cultivars

Durable disease resistance is a major but elusive goal of many crop improvement programs. Genomic approaches will have a significant impact on efforts to ameliorate plant diseases by increasing the definition of and access to gene pools available for crop improvement. This approach will involve the detailed characterization of the many genes that confer resistance, as well as technologies for the precise manipulation and deployment of resistance genes. Genomic studies on pathogens are providing an understanding of the molecular basis of specificity and the opportunity to select targets for more durable resistance. As goal of plant genomics is to understand the genetic and molecular basis of all biological processes in plants that are relevant to the species. This understanding is fundamental to allow efficient exploitation of plants as biological resources in the development of new cultivars with improved quality and reduced economic and environmental costs. This knowledge is also vital for the development of new plant diagnostic tools Traits considered of primary interest are, pathogen and abiotic stress resistance, quality traits for plant, and reproductive traits determining yield. A genome program can now be envisioned as a highly important tool for plant improvement. Such an approach to identify key genes and understand their function will result in a quantum leap in plant improvement. Additionally, the ability to examine gene expression will allow us to understand how plants respond to and interact with the physical environment and management practices. Bioinformatics plays several roles in breeding for disease resistance. It will be important for acquiring and organizing large amounts of information. It will also allow the visualization of information from heterogeneous datasets to facilitate the selection of superior individuals.

Disease resistance is only one of several traits under selection in a breeding program. Bioinformatics will therefore play an increasing role in integrating phenotypic and pedigree information for agronomic as well as resistance traits. Improved algorithms and increased computing power will make it possible to simulate and optimize selection strategies as well as to model the epidemiology of pathogens. According to Vassilev et al. the key role of bioinformatics for plant improvements may include: to encourage the submission of all sequence data into the public domain, through repositories, to provide rational annotation of genes, proteins and phenotypes, and to elaborate relationships both within the plants' data and between plants and other organisms to provide data including sequence information, information on mutations, markers, maps, functional discoveries, and other.

Production of Disease Free Planting Materials

Most agricultural biotechnologies involve tissue culture and DNA-based markers for germplasm conservation, production of disease-free planting material, and assistance to genetic improvement. More recently, Latin America countries such as, Argentina, Brazil, Colombia, Honduras, Mexico and Uruguay have commercially grown transgenic crops. Advanced biotechnologies, such as genetic sequencing and microarray genomics, are differentially utilized in some Latin America countries, with Brazil being at the forefront, for characterization, mapping, and trait screening for important crops and pathogens. With this regard the role of bioinformatics is indispensable.

Bioinformatics in Computer-Aided Drug Design

Computer-Aided Drug Design (CADD) is a specialized discipline that uses computational methods to simulate drug-receptor interactions. CADD methods are heavily dependent on bioinformatics tools, applications and databases. As such, there is considerable overlap in CADD research and bioinformatics.

There are several key areas where bioinformatics supports CADD research:

- Virtual High-Throughput Screening (vHTS): Pharmaceutical companies are always searching for new leads to develop into drug compounds. One search method is virtual high-throughput screening. In vHTS, protein targets are screened against databases of small-molecule compounds to see which molecules bind strongly to the target. If there is a "hit" with a particular compound, it can be extracted from the database for further testing. With today's computational resources, several million compounds can be screened in a few days on sufficiently large clustered computers. Pursuing a handful of promising leads for further development can save researchers considerable time and expense. ZINC is a good example of a vHTS compound library.

- Sequence Analysis: In CADD research, one often knows the genetic sequence of multiple organisms or the amino acid sequence of proteins from several species. It is very useful to determine how similar or dissimilar the organisms are based on gene or protein sequences. With this information one can infer the evolutionary relationships of the organisms, search for similar sequences in bioinformatic databases and find related species to those under investigation. There are many bioinformatic sequence analysis tools i×\5 that can be used to determine the level of sequence similarity.

- Homology Modeling: Another common challenge in CADD research is determining the 3-D structure of proteins. Most drug targets are proteins, so it's important to know their 3-D structure in detail. It's estimated that the human body has 500,000 to 1 million proteins. However, the 3-D structure is known for only a small fraction of these. Homology modeling is one method used to predict 3-D structure. In homology modeling, the amino acid sequence of a specific protein (target) is known, and the 3-D structures of proteins related to the target (templates) are known. Bioinformatics software tools are then used to predict the 3-D structure of the target based on the known 3-D structures of the templates. MODELLER is a well-known tool in homology modeling, and the SWISS-MODEL Repository is a database of protein structures created with homology modeling.

- Similarity Searches: A common activity in biopharmaceutical companies is the search for drug analogues. Starting with a promising drug molecule, one can search for chemical compounds with similar structure or properties to a known compound. There are a variety of methods used in these searches, including sequence similarity, 2D and 3D shape similarity, substructure similarity, electrostatic similarity and others. A variety of bioinformatic tools and search engines are available for this work.

- Drug Lead Optimization: When a promising lead candidate has been found in a drug discovery program, the next step (a very long and expensive step!) is to optimize the structure and properties of the potential drug. This usually involves a series of modifications to the primary structure (scaffold) and secondary structure (moieties) of the compound. This process can be enhanced using software tools that explore related compounds (bioisosteres) to the lead candidate. OpenEye's WABE is one such tool. Lead optimization tools such as WABE offer a rational approach to drug design that can reduce the time and expense of searching for related compounds.

- Physicochemical Modeling: Drug-receptor interactions occur on atomic scales. To form a deep understanding of how and why drug compounds bind to protein targets, we must consider the biochemical and biophysical properties of both the drug itself and its target at an atomic level. Swiss-PDB is an excellent tool for doing this. Swiss-PDB can predict key physicochemical properties, such as hydrophobicity and polarity that have a profound influence on how drugs bind to proteins.

- Drug Bioavailability and Bioactivity: Most drug candidates fail in Phase III clinical trials after many years of research and millions of dollars have been spent on them. And most fail because of toxicity or problems with metabolism. The key characteristics for drugs are Absorption, Distribution, Metabolism, Excretion, Toxicity (ADMET) and efficacy—in other words bioavailability and bioactivity. Although these properties are usually measured in the lab, they can also be predicted in advance with bioinformatics software.

Benefits of CADD

CADD methods and bioinformatics tools offer significant benefits for drug discovery programs.

- Cost Savings: The Tufts Report suggests that the cost of drug discovery and development has reached $800 million for each drug successfully brought to market. Many biopharmaceutical companies now use computational methods and bioinformatics tools to reduce this cost burden. Virtual screening, lead optimization and predictions of bioavailability and bioactivity can help guide experimental research. Only the most promising experimental lines of inquiry can be followed and experimental dead-ends can be avoided early based on the results of CADD simulations.

- Time-to-Market: The predictive power of CADD can help drug research programs choose only the most promising drug candidates. By focusing drug research on specific lead candidates and avoiding potential "dead-end" compounds, biopharmaceutical companies can get drugs to market more quickly.

- Insight: One of the non-quantifiable benefits of CADD and the use of bioinformatics tools is the deep insight that researchers acquire about drug-receptor interactions. Molecular models of drug compounds can reveal intricate, atomic scale binding properties that are difficult to envision in any other way. When we show researchers new molecular models of their putative drug compounds, their protein targets and how the two bind together, they often come up with new ideas on how to modify the drug compounds for improved fit. This is an intangible benefit that can help design research programs.

Chapter 4
Gene and DNA Sequence Analysis

Sequence analysis seeks to understand the structure, feature and function of a DNA, RNA or peptide sequence using a broad range of analytical methods. Gene prediction, DNA sequencing and whole genome sequencing are some of the fields of study within this discipline. The chapter closely examines these key concepts of gene and DNA sequence analysis to provide an extensive understanding of the subject.

Gene Prediction and Methods

Gene prediction is the process of inferring the sequence of the functional products encoded in genomic DNA sequences.

Gene Prediction Methods

There are two basic problems in gene prediction: prediction of protein coding regions and prediction of the functional sites of genes. A large number of researches working on this subject have accumulated, which can be classified into four generations in summary. The first generation of programs was designed to identify approximate locations of coding regions in genomic DNA. The most widely known programs were probably TestCode and GRAIL. But they could not accurately predict precise exon locations. The second generation, such as SORFIND and Xpound, combined splice signal and coding region identification to predict potential exons, but did not attempt to assemble predicted exons into complete genes. The next generation of programs attempted the more difficult task of predicting complete gene structures. A variety of programs have been developed, including GeneID GeneParser, GenLang, and FGENEH, However, the performance of those programs remained rather poor. Moreover, those programs were all based on the assumption that the input sequence contains exactly one complete gene, which is not often the case. To solve this problem and improve accuracy and applicability further, GENSCAN and AUGUSTUS were developed, which could be classified into the fourth generation.

There are mainly two classes of methods for computational gene prediction. One is based on sequence similarity searches, while the other is gene structure and signal-based searches, which is also referred to as ab initio gene finding.

Sequence Similarity Searches

Sequence similarity search is a conceptually simple approach that is based on finding similarity in gene sequences between ESTs (expressed sequence tags), proteins, or other genomes to the input genome. This approach is based on the assumption that functional regions (exons) are more conserved evolutionarily than nonfunctional regions (intergenic or intronic regions). Once there is similarity between a certain genomic region and an EST, DNA, or protein, the similarity information can be used to infer gene structure or function of that region. EST-based sequence similarity usually has drawbacks in that ESTs only correspond to small portions of the gene sequence, which means that it is often difficult to predict the complete gene structure of a given region.

Local alignment and global alignment are two methods based on similarity searches. The most common local alignment tool is the BLAST family of programs, which detects sequence similarity to known genes, proteins, or ESTs. Two more types of software, PROCRUSTES and GeneWise, use global alignment of a homologous protein to translated ORFs in a genomic sequence for gene prediction. A new heuristic method based on pairwise genome comparison has been implemented in the software called CSTfinder. The biggest limitation to this type of approaches is that only about half of the genes being discovered have significant homology to genes in the databases.

Ab Initio Gene Prediction Methods

The second class of methods for the computational identification of genes is to use gene structure as a template to detect genes, which is also called ab initio prediction. Ab initio gene predictions rely on two types of sequence information: signal sensors and content sensors. Signal sensors refer to short sequence motifs, such as splice sites, branch points, polypyrimidine tracts, start codons and stop codons. Exon detection must rely on the content sensors, which refer to the patterns of codon usage that are unique to a species, and allow coding sequences to be distinguished from the surrounding non-coding sequences by statistical detection algorithms.

Many algorithms are applied for modeling gene structure, such as Dynamic Programming, linear discriminant analysis, Linguist methods, Hidden Markov Model and Neural Network. Based on these models, a great number of ab initio gene prediction programs have been developed. Some of the frequently used ones are shown in Table, among which the programs GeneParser, Genie and GRAIL combine similarity searches.

Table· *Ab initio* Gene Prediction Programs (Possibly with Homology Integration).

Program	Organism	Algorithm	Homology
GeneID	Vertebrates, plants	DP	
FGENESH	Human, mouse, Drosophila, rice	HMM	
GeneParser	Vertebrates	NN	EST
Genie	Drosophila, human, other	GHMM	protein
GenLang	Vertebrates, Drosophila, dicots	Grammar rule	
GENSCAN	Vertebrates, Arabidopsis, maize	GHMM	
GlimmerM	Small eukaryotes, Arabidopsis, rice	IMM	
GRAIL	Human, mouse, Arabidopsis, Drosophila	NN, DP	EST, cDNA
HMMgene	Vertebrates, *C. elegans*	CHMM	
AUGUSTUS	Human, Arabidopsis	IMM,WWAM	
MZEF	Human, mouse, Arabidopsis, Fission yeast	Quadratic discriminant analysis	

The most successful programs so far are based on Hidden Markov Model. Readers interested in other algorithms can learn from references. In Hidden Markov Model, transitions between sub-models corresponding to particular gene components are modeled as unobserved ("hidden") Markov processes, which determine the probability of generating particular (observable) nucleotides. Since exon and intron lengths appear to be constrained by factors related to pre-mRNA splicing, and do not exhibit geometric distributions, a more general model is required to accurately account for the lengths of exons and introns in real genes. So a Generalized Hidden Markov Model (GHMM) is developed, in which subsequent states are generated according to a Markov chain but have arbitrary (instead of fixed unit) length distributions. Figure illustrates the state transition in eukaryotic genomic sequences.

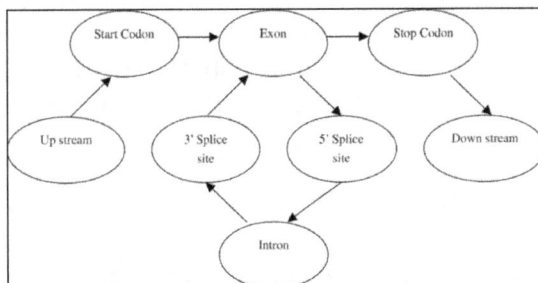

Figure: State transition of HMM modeling eukaryotic genes.

Suppose we are given a DNA sequence S of length L and a parse ϕ also of length L. The conditional probability of the parse ϕ, given that the sequence generated is S, can be computed using Bayes' Rule as:

$$P\{\phi|S\} = \frac{P\{\phi,S\}}{\sum\limits_{\psi \in \phi L} P\{\psi,S\}}$$

Here, ϕL is the set of all parses of length L. Now, given a particular DNA sequence S, we can find a parse ϕL that maximizes the likelihood of generating. In other words, for a particular sequence, we can find the functional unit (for example, the promoter region) that the sequence is most likely to represent. Thus, the model can be used for automatic annotation of DNA sequences.

Other Methods

The major limitation with HMM method is that we have a little knowledge of gene structures, especially for new sequencing genomes. Furthermore, current set of known genes is limited and certainly does not represent all potential gene features or their organizational themes. So recently some techniques in physics and signal processing have been applied to recognize genes.

It is well known that base sequences in the protein-coding regions of DNA molecules have a period-3 component because of the codon structure involved in the translation of base sequences into amino acids. Discrete Fourier Transform (DFT) is suitable for processing periodicity. For a DNA sequence of length N, assume $u_A(n)$, $u_T(n)$, $u_C(n)$, and $u_G(n)$, which represent the binary indicator function for the corresponding nucleotide. It takes the value 1 at index n if the corresponding nucleotide is present at that position, and takes the value 0 otherwise. Applying DFT to each of these sequences produces four spectral representations, represented as $U_A(k)$, $U_T(k)$, $U_C(k)$, and $U_G(k)$, respectively. The total frequency spectrum of the given DNA sequence is defined as,

$$S(k) = |U_A(k)|^2 + |U_T(k)|^2 + |U_C(k)|^2 + |U_G(k)|^2$$

In coding regions of DNA, $S(k)$ typically has a peak at the frequency $k = N/3$, whereas in noncoding regions, it generally does not have any significant peaks. By this property, gene predictor can be constructed. In 2003, a new measure for gene prediction in eukaryotes was presented by Kotlar and Lavner , which was based on DFT. The phase of the DFT at a frequency of 1/3 was distributed with a bell-shaped curve around a central value in coding regions, whereas in noncoding regions, the distribution was close to uniform. This regularity was used for discriminating between coding and noncoding regions in a given nonannotated genomic sequence.

The Z curve method is another powerful tool in visualizing and analyzing DNA sequences. It has been applied to recognize coding sequences in the human genome, and to find genes in the genomes of yeast and Vibrio cholerae. For predicting short coding sequence, it shows higher accuracy than GENSCAN, which is considered as one of the best ab initio gene prediction programs, while it is much simpler computationally than the latter.

In addition, with many genome sequencing projects currently under way, the comparative genome approach is becoming more promising in the field of gene prediction. In practice, its performance will depend on the evolutionary distance between the compared sequences. Initial results show that the relationship is not straightforward. Indeed, a greater evolutionary distance allows some algorithms to more accurately discriminate between coding and non-coding sequence conservation. Such comparative genome programs are often computer intensive and consequently much work remains to be done.

Evaluation of Gene Prediction Programs

The abundance of gene prediction program raises the problem of adequate evaluation of prediction program quality. Comparison of the accuracy and reliability must take into account the type of algorithms, for example, neural network, Hidden Markov Model, or others; the number of sequences used for training and testing; and the method used for evaluation. It is impossible to rank the predictors by only a single measure.

Sensitivity (Sn) and Specificity (Sp) are probably the two most widely used measures, which are explained by Burset and Guigó in detail. The accuracy of the predictions can be measured at three different levels: coding nucleotide sequence, exonic structure, and protein product. The nucleotide level accuracy that measures Sn, Sp, CC (correlation coefficient) and AC (approximate coefficient) gives an overall sense of how closely the predicted and actual coding regions are in a sequence alignment, but does not accurately reflect the identification of precise exon boundaries. Evaluation at the exon level mainly provides how well the sequence signals (splice sites, start codon, and stop codon, etc.) are identified. The accuracy can be measured by comparing predicted and real exons along the test sequences.

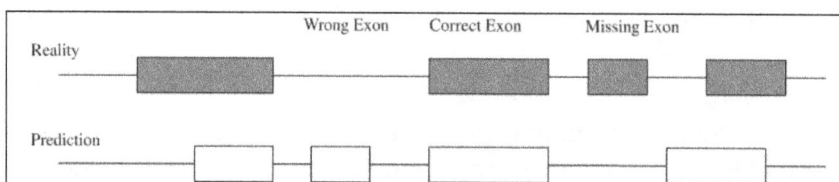

Figure: Evaluation of gene prediction accuracy at the exon level.

Thus, sensitivity (Sn), specificity (Sp), miss rate (MR) and wrong rate (WR) are expressed as following:

$$Sn = \frac{CE}{AE} \quad Sp = \frac{CE}{PE}$$

$$MR = \frac{ME}{AE} \quad WR = \frac{WE}{PE}$$

(AE, actural exons; PE, predicted exons; CE, correct exons; WE, wrong exons; ME, missing exons).

At the protein level, the accuracy is measured by comparing the protein product encoded by the actual gene in the test sequence with the protein product encoded by the predicted gene. In the gene prediction literatures, only Fields and Soderlund provided an evaluation of the gm program at the final protein product level, which indicated that it is not widely used.

The prediction accuracy of some usual programs has been tested on Burset and Guigó's sequence set , and the results at exon level are illustrated in . It shows that GENSCAN based on GHMM is significantly more accurate than other programs.

Table: Accuracy Comparisons of Gene Prediction Programs

Program	Sn	Sp	MR	WR
GENSCAN	0.78	0.81	0.09	0.05
FGENEH	0.61	0.64	0.15	0.12
GeneID	0.44	0.46	0.28	0.24
Genie	0.55	0.48	0.17	0.33
GenLang	0.51	0.52	0.21	0.22
GeneParser2	0.35	0.40	0.34	0.17
GRAIL2	0.36	0.43	0.25	0.11
SORFIND	0.42	0.47	0.24	0.14
Xpound	0.15	0.18	0.33	0.13

Since the early eighties of the twentieth century, there has been great progress in the development of computational gene prediction. However, some problems have not yet been solved. First, short exons are difficult to locate, because discriminative statistical characteristics are less likely to appear in short sequences. The more difficult cases are those where the length of a coding exon is a multiple of three (typically 3, 6 or 9 bp), because missing such exons will not cause a problem in the exon assembly as they do not introduce any changes in the frame. Lately, Gao and Zhang (26) compared the performance of various algorithms for recognizing short coding sequences and validated that the Z curve method is the best one.

Second, the problem of alternatively splicing has not yet been solved effectively, which in particular is an important regulatory mechanism in higher eukaryotes. Some gene prediction programs tried to handle this through the identification of sub-optimal exons (GENSCAN and MZEF). Nevertheless, a more relevant approach would consist of improving the identification of the intronic and exonic signals that dictate the choice of alternatively splicing sites.

In addition, the evaluation system of gene prediction programs is still in need of improvement. Some of the measures mentioned above often give results contradictory to each other, because many of them emphasize only a few or even only one of the several aspects of the prediction quality. So more reasonable and comprehensive criterions are needed for evaluation of gene prediction programs. Recently, Bajic introduced averaged score measure (ASM) and used it to assess the quality of programs for eukaryotic promoter prediction.

Further more, in order to compensate the insufficiency of any individual gene prediction program, the computational method to construct gene models by multiple evidences is becoming more promising. For the nonannotated genomic sequences, a diverse set of sources can be combined for annotation, including the locations of gene predictions from ab initio gene finders, protein sequence alignments, ESTs and cDNA alignments, promoter predictions, splice site predictions, and so on. Such integrative approach has been proved to consistently outperform even the best individual gene finder, and in some cases, can produce dramatic improvements in sensitivity and specificity.

Finally, it should be emphasized that for all gene prediction methods, the performances depend on the current biological knowledge to a large extent, especially knowledge at the molecular level of gene expression. So it requires great efforts by both experimental and computational biologists to make gene prediction more accurate, which can definitely speed up gene discovery and knowledge mining.

DNA Sequencing

DNA sequencing is the technique used to determine the nucleotide sequence of DNA (deoxyribonucleic acid). The nucleotide sequence is the most fundamental level of knowledge of a gene or genome. It is the blueprint that contains the instructions for building an organism, and no understanding of genetic function or evolution could be complete without obtaining this information.

First-Generation Sequencing Technology

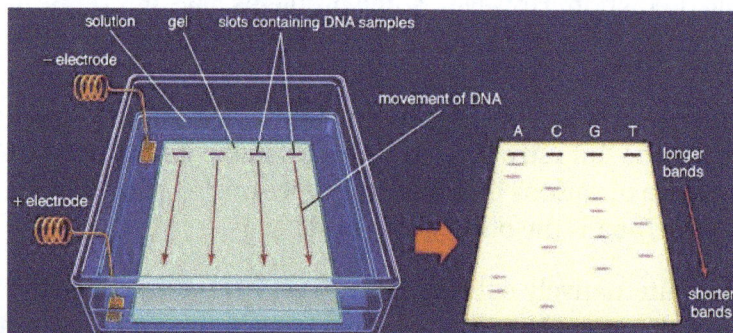

In gel electrophoresis an electric field is applied to a buffer solution covering an agarose gel, which has slots at one end containing DNA samples. The negatively charged DNA molecules travel through the gel toward a positive electrode and are separated based on size as they advance.

So-called first-generation sequencing technologies, which emerged in the 1970s, included the Maxam-Gilbert method, discovered by and named for American molecular biologists Allan M. Maxam and Walter Gilbert, and the Sanger method (or dideoxy method), discovered by English biochemist Frederick Sanger. In the Sanger method, which became the more commonly employed of the two approaches, DNA chains were synthesized on a template strand, but chain growth was stopped when one of four possible dideoxy nucleotides, which lack a 3' hydroxyl group, became incorporated, thereby preventing the addition of another nucleotide. A population of nested, truncated DNA molecules was produced that represented each of the sites of that particular nucleotide in the template DNA. The molecules were separated according to size in a procedure called electrophoresis, and the inferred nucleotide sequence was deduced by a computer. Later, the method was performed by using automated sequencing machines, in which the truncated DNA molecules, labeled with fluorescent tags, were separated by size within thin glass capillaries and detected by laser excitation.

Next-Generation Sequencing Technology

Next-generation (massively parallel, or second-generation) sequencing technologies have largely supplanted first-generation technologies. These newer approaches enable many DNA fragments (sometimes on the order of millions of fragments) to be sequenced at one time and are more cost-efficient and much faster than first-generation technologies. The utility of next-generation technologies was improved significantly by advances in bioinformatics that allowed for increased data storage and facilitated the analysis and manipulation of very large data sets, often in the gigabase range (1 gigabase = 1,000,000,000 base pairs of DNA).

Applications of DNA Sequencing Technologies

Knowledge of the sequence of a DNA segment has many uses. First, it can be used to find genes, segments of DNA that code for a specific protein or phenotype. If a region of DNA has been sequenced, it can be screened for characteristic features of genes. For example, open reading frames (ORFs)—long sequences that begin with a start codon (three adjacent nucleotides; the sequence of a codon dictates amino acid production) and are uninterrupted by stop codons (except for one at their termination)—suggest a protein-coding region. Also, human genes are generally adjacent to so-called CpG islands—clusters of cytosine and guanine, two of the nucleotides that make up DNA. If a gene with a known phenotype (such as a disease gene in humans) is known to be in the chromosomal region sequenced, then unassigned genes in the region will become candidates for that function. Second, homologous DNA sequences of different organisms can be compared in order to plot evolutionary relationships both within and between species. Third, a gene sequence can be screened for functional regions. In order to determine the function of a gene, various domains can be identified that are common to proteins of similar function. For example, certain amino acid sequences within a gene are always found in proteins that span a cell membrane; such amino acid stretches are called transmembrane domains. If a transmembrane domain is found in a gene of unknown function, it suggests that the encoded protein is located in the cellular membrane. Other domains characterize DNA-binding proteins. Several public databases of DNA sequences are available for analysis by any interested individual.

The applications of next-generation sequencing technologies are vast, owing to their relatively low cost and large-scale high-throughput capacity. Using these technologies, scientists have been able to rapidly sequence entire genomes (whole genome sequencing) of organisms, to discover genes involved in disease, and to better understand genomic structure and diversity among species generally.

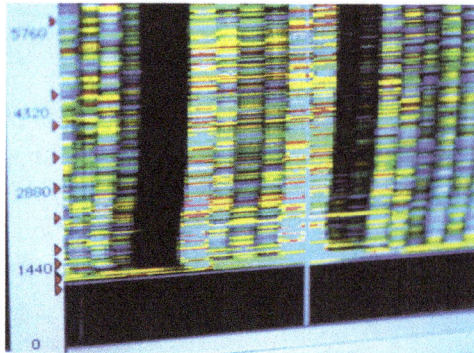

DNA sequencing A nucleotide sequence determined using DNA sequencing technologies.

DNA Sequence Data Analysis

DNA sequencing is used to determine the sequence of individual genes, full chromosomes or entire genomes of an organism. DNA sequencing has also become the most efficient way to sequence RNA or proteins.

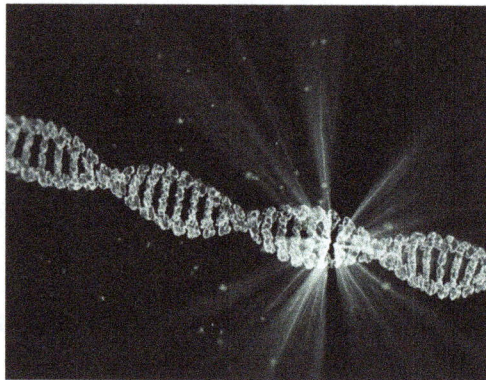

DNA Molecule

Whole Genome Sequencing and Sequence Assembly

A DNA sequencing reaction produces a sequence that is several hundred bases long. Gene sequences are typically thousands of bases long. The largest known gene is the one associated with Duchenne muscular dystrophy. It is approximately 2.4 million bases in length. In order to study one whole gene, scientists use a simple strategy known as shotgun sequencing. The long DNA sequence is assembled from a series of shorter overlapping sequences. Let's see what happens in the shotgun sequencing approach.

Shotgun Sequencing

Special machines, known as sequencing machines are used to extract short random DNA sequences from a particular genome we wish to determine (target genome). Current DNA sequencing technologies cannot read one whole genome at once. It reads small pieces of between 20 and 30000 bases, depending on the technology used. These short pieces are called reads. Special software are used to assemble these reads according to how they overlap, in order to generate continuous strings called contigs. These contigs can be the whole target genome itself, or parts of the genome.

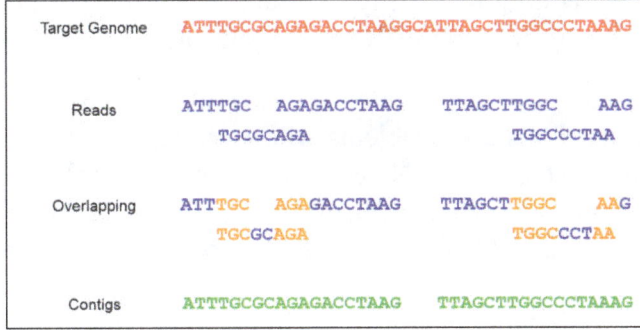

Shotgun Sequencing

The process of aligning and merging fragments from a longer DNA sequence, in order to reconstruct the original sequence is known as Sequence Assembly.

In order to obtain the whole genome sequence, we may need to generate more and more random reads, until the contigs match to the target genome.

Sequence Assembly Problem

The sequence assembly problem can be described as follows:

- Given a set of sequences, find the minimal length string containing all members of the set as substrings.

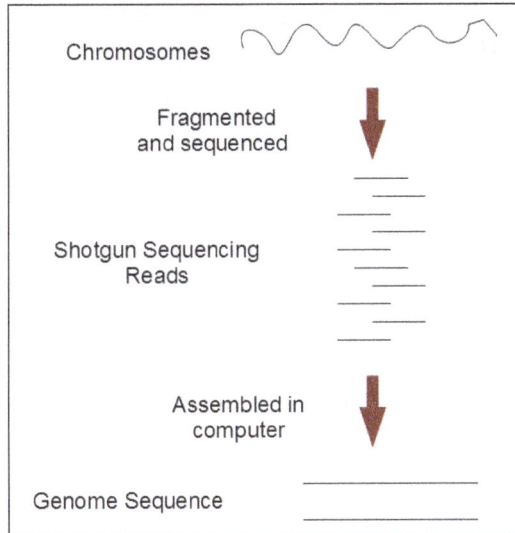

- This problem is further complicated due to the existence of repetitive sequences in the genome as well as substitutions or mutations withing them.

- The sequence assembly problem can be compared to a real life scenario.

Assume that you take many copies of a book, pass each of them through a shredder with a different cutter, and then you try to make the text of the book back together just by gluing together the shredded pieces. It is obvious that this task is pretty difficult. Furthermore, there are some extra practical issues as well. The original copy may have many repeated paragraphs, and some shreds may be modified during shredding to have typos. Parts from another book may have also been added in, and some shreds may be completely unrecognizable.

It sounds very confusing and quite impossible to be carried out. This problem is known to be NP Complete. NP complete problems are problems whose status is unknown. No polynomial time algorithm has yet been discovered for any NP complete problem, nor has anybody yet been able to prove that no polynomial-time algorithm exists for any of them. However, there are greedy algorithms to solve the sequence assembly problem, where experiments have proven to perform fairly well in practice.

A common method used to solve the sequence assembly problem and perform sequence data analysis is sequence alignment.

Sequence Alignment

Sequence alignment is a method of arranging sequences of DNA, RNA, or protein to identify regions of similarity. The similarity being identified, may be a result of functional, structural, or evolutionary relationships between the sequences.

If we compare two sequences, it is known as pairwise sequence alignment. If we compare more than two sequences, it is known as multiple sequence alignment.

Next-Generation Sequencing

Next-generation sequencing (NGS), also known as high-throughput sequencing, is the collective term used to describe many different modern sequencing technologies such as,

- Illumina (Solexa) sequencing
- Roche 454 sequencing
- Ion torrent Proton / PGM sequencing
- Solid sequencing

These recent technologies allow us to sequence DNA and RNA much more quickly and cheaply than the previously used Sanger sequencing, and have revolutionized the study of genomics.

Nucleic Acid Amplification

Nucleic acid amplification is a valuable molecular tool not only in basic research but also in application oriented fields, such as clinical medicine development, infectious diseases diagnosis, gene cloning and industrial quality control. PCR was the first nucleic acid amplification method. With the advancement of research, a no of alternative nucleic acid amplification methods has been developed such as loop mediated isothermal amplification, nucleic acid sequence based amplification, strand displacement amplification, multiple displacement amplification. Most of the alternative methods are isothermal obviating the need for thermal cyclers. Though principles of most of the alternate methods are relatively complex than that of PCR, they offer better applicability and sensitivity in cases where PCR has limitations. Most of the alternate methods still have to prove themselves through extensive validation studies and are not available in commercial form; they pose the potentiality to be used as replacements of PCR. Continuous research is going on in different parts of the world to make these methods viable technically and economically.

Nucleic acid amplification is a pivotal process in biotechnology and molecular biology and has been widely used in research, medicine, agriculture and forensics. Polymerase chain reaction (PCR) was the first nucleic acid amplification method developed and until now has been the method of choice since its invention by Mullis. PCR is the preferred method for application oriented fields involving nucleic acid amplification for its simplicity, easier methodology, extensively validated standard operating procedure and availability of reagents and equipments. However, PCR has a good no of limitations, including high cost of equipment, contamination chances, sensitivity to certain classes of contaminants and inhibitors, requirement of thermal cycling etc. These limitations gave birth to alternative methods such as loop mediated isothermal amplification (LAMP), nucleic acid sequence based amplification (NASBA), self-sustained sequence replication (3SR), rolling circle amplification (RCA) etc., most of which are isothermal nucleic acid amplification methods obviating the need of thermal cycler. These methods offer potential advantages over PCR for speed, cost, scale or portability.

Alternative Methods of Polymerase Chain Reaction

Several alternative amplification methods have been developed already, such as LAMP, 3SR or NASBA, strand displacement amplification (SDA) and RCA, ligase chain reaction (LCR).

Loop Mediated Isothermal Amplification

LAMP is a specific, simple, rapid and cost-effective isothermal nucleic acid amplification method. LAMP has an improved simple visual amplicon detection system. LAMP relies on the auto-cycling strand displacement deoxyribonucleic acid (DNA) synthesis which is carried out at 60-65°C for 45-60 min in the presence of Bacillus stearothermophylus (Bst) DNA polymerase, deoxyribonucleotide triphosphate (dNTPs), specific primers and the target DNA template. The LAMP method employs a DNA polymerase with high strand displacement activity and a set of four specially constructed primers (two inner and two outer primer) that recognize six distinct sequences on the target DNA. The mechanism of the LAMP amplification reaction includes three steps: Production of starting material, cycling amplification and elongation and recycling. High-level of precision can be attained without expensive equipments. There are fewer and simpler sample preparation steps compared with conventional PCR and real-time PCR. Substantial alteration of the fluorescence of the reaction tube can be visualized without costly specialized equipment as the signal recognition system is highly sensitive. LAMP is a one-step amplification reaction taking only 30-60 min. LAMP is more resistant to various inhibitory compounds present in clinical samples than PCR, so there is no need for extensive DNA purification. By combination with reverse transcription (RT),

LAMP can amplify ribonucleic acid (RNA) sequences with high efficiency. It is highly sensitive and able to detect DNA at as few as six copies in the reaction mixture. LAMP has the potential to be helpful in basic research on medicine and pharmacy, environmental hygiene, point-of-care testing and cost-effective diagnosis of infectious diseases. LAMP is as suitable for DNA sequencing as PCR, in terms of both Sanger sequencing and Pyrosequencing.

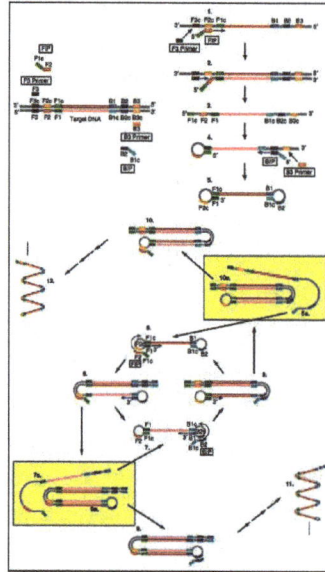

Figure: Schematic description of loop mediated isothermal amplification assay

Nucleic Acid Sequence based Amplification

NASBA, also known as 3SR and transcription mediated amplification, is an isothermal transcription-based amplification system. NASBA specifically designed for the detection of RNA targets. In some NASBA systems, DNA can also be amplified. The complete amplification reaction is performed at the predefined temperature of 41 °C. Throughout the amplification reaction, constant temperature is maintained allowing each step of the reaction to proceed as soon as amplification intermediate formed. The exponential kinetic of the NASBA process is attributed by multiple transcription of RNA copies from a given DNA product, is intrinsically more efficient than DNA-amplification methods limited to binary increases per cycle. This amplification system uses a consortium of three enzymes (avian myeloblastosis virus reverse transcriptase, RNase H and T7 DNA dependent RNA polymerase) leading to main amplification product of single-stranded RNA. NASBA RNA product can be sequenced directly with a dideoxy method using RT and a labeled oligonucleotide primer. The length of the target sequence to be amplified efficiently is limited to approx 100-250 nucleotides. High-level of precision can be acquired without expensive equipments. NASBA amplicon detection step has significantly improved, incorporation of the use of enzyme-linked gel assay, enzymatic bead-based detection and electrochemiluminescent (ECL) detection, molecular beacon technology and fluorescent correlation spectroscopy. In clinical use and pathogen detection, NASBA pose theoretically higher analytical sensitivity than reverse transcription-polymerase chain reaction RT-PCR making it an established diagnostic tool. It has potential for detection and differentiation of viable cells through specific and sensitive amplification of messenger RNA, even against the background of genomic DNA.

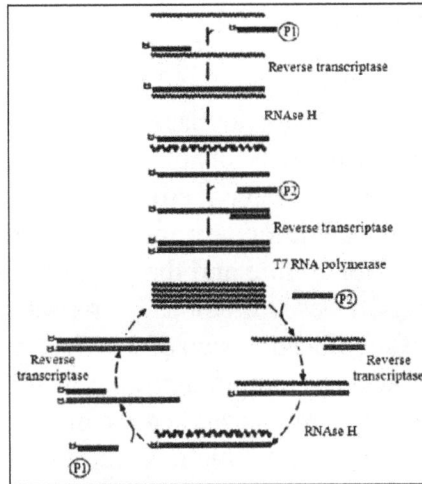

Figure: Principles of nucleic acid sequence based amplification

Strand Displacement Amplification

SDA, first described in 1992, is an isothermic amplification method, which utilizes four different primers of which a primer containing a restriction site (a recognition sequence for HincII exonuclease) is annealed to the DNA template. An exonuclease-deficient fragment of Eschericia coli DNA polymerase 1 (exo-Klenow) elongates the primers. Each SDA cycle consists of primer binding to a displaced target fragment, extension of the primer/target complex by exoklenow, nicking of the resultant hemiphosphothioate HincII site, dissociation of HincII from the nicked site and extension of the nick and displacement of the downstream strand by exo-Klenow. This method can be performed at high temperatures. In a single reaction, 109 copies of target DNA can be produced in less than an hour. As with the other target amplification technologies (PCR, LCR, 3SR), only semi-quantitation is possible by this method. A major limitation of SDA is its inability to efficiently amplify long target sequences. SDA is the basis for some commercial detection tests such as BDProbeTec (Becton Dickinson, Franklin Lakes, NJ, USA) and has been evaluated recently for the identification of Mycobacterium tuberculosis directly from clinical specimens. The efficiency and robustness of this technology still remains to be proven in the large clinical studies. A recent development has been reported for real-time sequence specific DNA target detection using the fluorogenic reporter probes.

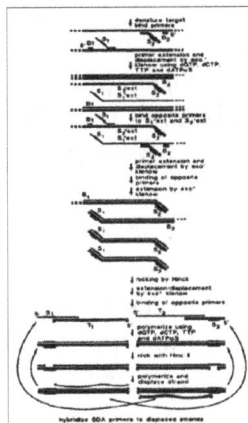

Figure: Target generation scheme for strand displacement amplification

Multiple Displacement Amplification

The MDA is an isothermal, strand-displacing method based on the use of the highly processive and strand-displacing DNA polymerase from bacteriophage Ø29, in conjunction with modified random primers to amplify the entire genome with high-fidelity. It has been developed to amplify all DNA in a sample from a very small amount of starting material. MDA by Ø29 DNA polymerase involves incubating Ø29 DNA polymerase, dNTPs, random hexamers and denatured template DNA at 30 °C for 16-18 h. The enzyme is inactivated at 65 °C for 10 min and the product DNA can be used directly in downstream applications In contrast to PCR-based methods no repeated cycling is required, but a short initial denaturation followed by the amplification step of 6-18 h and a final inactivation of the enzyme is needed. This method can also be used to generate capture probes for microarrays, to produce highly pure DNA or even to amplify stored DNA. MDA could be the method of choice when limited amounts of sample are available. Sensitivity and yields are high: About 20-30 µg of DNA can be obtained from as few as 1-10 copies of human genomic DNA. In addition, MDA can be carried out directly from biological samples including crude whole blood and tissue culture cells. MDA method presents several characteristics that, in combination with specific hybridization to macroarray and microarray, may be suitable for multiplex qualitative detection and identification. The utility of MDA has not been fully assessed for use in applications such as sample archiving, forensics and single cell clinical diagnostics.

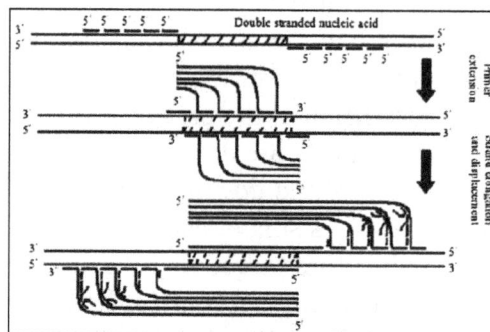

Figure: Schematic representation of multiple displacement amplification mechanism

Rolling Circle Amplification

RCA is an isothermal nucleic acid amplification method. RCA technology enables the amplification of the probe DNA sequences more than 10^9 fold both in solution and on the solid phase at a single temperature. It has the ability to readily detect down to a few target-specific circularized probes in a test sample. In RCA reaction, numerous rounds of isothermal enzymatic synthesis is involved. Ø29 DNA polymerase extends a circle-hybridized primer by continuously progressing around the circular DNA probe of several dozen nucleotides to replicate its sequence over and over again. A major advantage of RCA is that unlike PCR, this technology is resistant to contamination and unlike some other isothermal technologies, requires little or no assay optimization. The capacity of RCA to yield the surface-bound amplification products offers significant advantages to in situ or microarray hybridization assays. In linear RCA, the product of amplification remains tethered to the target molecule. RCA seems well-suited to cell-and tissue-based assays in conjunction with the isothermal nature of the RCA reaction and the capability to localize multiple markers simultaneously. RCA is also suitable in cases where it is critical to maintain morphological information. RCA amplification permits the localization of signals, thus representing single molecules with specific genetic traits or biochemical features.

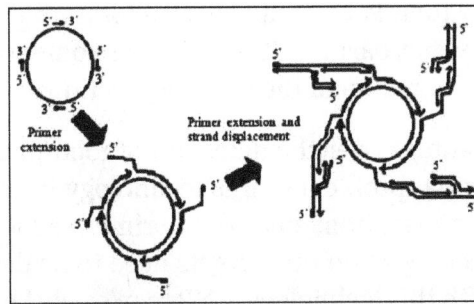

Figure: Scheme for multiply-primed rolling circle amplification

RCA reactions exhibit an excellent sequence specificity that is favorable for genotyping or mutation detection and allows to unambiguously identifying DNA markers on the excessive unrelated background. As compared with PCR, the RCA based DNA diagnostics are capable of higher multiplexity and they are less prone to amplification errors, thus allowing contamination-resistant detection of target molecules in a variety of testing formats. RCA has the potential for a highly localized isothermal detection of the designated sites on essentially intact duplex DNA with even superior specificity. The simplicity and efficiency of RCA technology, along with ease and accuracy of quantitation, makes it amenable for miniaturization and automation in the high-throughput analysis.

Ligase Chain Reaction

LCR is a cyclic DNA template-dependent amplification reaction. The method of DNA amplification is similar to PCR; however, LCR amplifies the probe molecule rather than producing amplicon through polymerization of nucleotides. LCR uses both a DNA polymerase enzyme and a DNA ligase enzyme to drive the reaction. LCR uses two complementary pairs of oligonucleotides that hybridize in close proximity on the target fragment. Only when the oligonucleotides correctly hybridize to the target sequence, the remaining nick between the oligonucleotides is ligated by a DNA ligase and a fragment equating to the total sequence of both oligonucleotides is generated. Once the probes have been ligated, the ligation product can serve as a template for annealing and future ligation. Like PCR, LCR requires a thermal circler to drive the reaction and each cycle results in a doubling of the target nucleic acid molecule. The detection of LCR products can be performed by electrophoresis or by an enzyme-linked immunosorbent assay (ELISA) like microplate procedure or real-time detection.

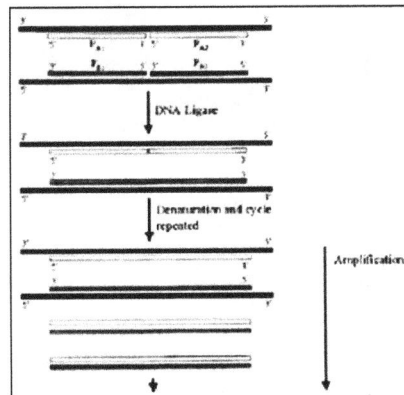

Figure: Ligase chain reaction

LCR can have greater specificity than PCR. It can be used for multiplex reactions rendering it suitable for detection of products by microarrays. Its limitations come from the specificity of the ligase reaction, which is restricted to the region of the ligation junction.

This technique has one disadvantage over the detection of food pathogen that it can detect DNA from dead organism. The potential problem of this technology in addition to the risk of contamination is clearly the lack of conformation, since only primer sequences are amplified. However, LDR sensitivity is limited, especially when rare targets need to be detected in the presence of high wild-type DNA background. With this technology, SNPs can easily be differentiated. It has been exploited to detect pathogens such as Neisseria gonorrhoneae, Chlamydia trachomatis and M. tuberculosis in human clinical specimens as well as detecting point mutations in C. trachomatis plasmid DNA, human immunodeficiency virus cloned fragments and in human sickle cell clinical samples.

Helicase Dependant Amplification

HDA is an isothermal nucleic acid amplification method using the replication fork mechanism. Basic principle of HDA is the unwinding activity of a DNA helicase in the presence of adenosine tri phosphate. Helicase separates the two strands of a DNA duplex generating single-stranded templates for the purpose of in vitro amplification of a target nucleic acid and the displaced DNA strands are coated by single-stranded binding proteins. Two-sequence specific primers anneal to the 3'- end of each single-stranded DNA (ssDNA) template and exonuclease-deficient DNA polymerases produce double-stranded DNA by extending the primers annealed to the target DNA. Exponential amplification can be achieved if this process repeats itself at a single temperature. This process allows multiple cycles of replication to be performed at a single incubation temperature, completely eliminating the need for thermo cycling equipment. The HDA amplicons can be detected using gel electrophoresis, real-time format and ELISA. High speed (100 bp/s) and processivity (10 kb/binding) are the major advantages of HDA. HDA offers several advantages over other isothermal DNA amplification methods. It has a simple reaction scheme and can be performed at a single temperature for the entire process. These properties offer a great potential for the development of simple portable DNA diagnostic devices to be used in the field and at the point-of-care.

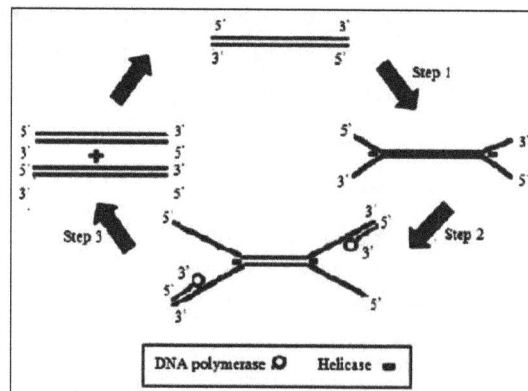

Figure: Helicase-dependent amplification process. Step 1: The helicase unwinds deoxyribonucleic acid (DNA) duplexes. Step 2: The primers anneal to the single-stranded DNA. Step 3: The primers extended by DNA polymerase; one duplex is amplified and converted to two duplexes. The double-stranded DNAs are separated by helicase and this chain reaction repeats itself.

Ramification Amplification Method

RAM is a novel isothermal nucleic acid amplification method. This technique is termed as RAM because the amplification power is derived from primer extension, strand displacement and multiple ramification (branching) points. This method uses a specially designed circular probe (C-probe) in which the 3' and 5' ends are brought together in juxtaposition by hybridization to a target. The two ends are then covalently linked by a T4 DNA ligase in a target-dependent manner, producing a closed DNA circle. In the presence of an excess of primers (forward and reverse primers), bacteriophage Ø29 DNA polymerase extends the bound forward primer along the C-probe and displaces the downstream strand, thus generating a multimeric ssDNA by continuously rolling over the closed circular DNA, analogous to the "rolling circle" replication of bacteriophages in vivo.

The multimeric ssDNA generated then serves as a template where multiple reverse primers hybridize, extend and displace downstream DNA and generate a large ramified (branching) DNA complex. This ramification process continues until all ssDNAs become double-stranded, resulting in an exponential amplification that distinguishes itself from the previously described non-exponential RCA. By using a unique bacteriophage DNA polymerase, Ø29 DNA polymerase, that has an intrinsic high processivity, it is possible to achieve significant amplification within 1 h at 35 °C. As RAM is an isothermal amplification method and large multimeric products are generated, cell morphological characteristics are preserved while amplification products are better localized in the cells, making this method ideal for in situ amplification. The RAM assay offers several advantages over other amplification techniques: (1) the primers readily bind to ssDNAs displaced by the DNA polymerase, enabling the reaction to be performed under isothermal conditions, obviating the need for a thermocycler; (2) generic primers amplify all probes with equal efficiency, resulting in better multiplex capability than conventional PCR; both ends of the probe can be ligated regardless of the nature of target (DNA or RNA), eliminating the need for RT for detecting RNA and creating a uniform assay format for both RNA and DNA detection; and (4) ligation requires that both probe termini hybridize with perfect matching, permitting the detection of a single-nucleotide polymorphism. It can be readily used in clinical laboratories for the detection of genes and infectious agents in various areas, such as Hematology, Oncology, infectious disease, Pathology, forensics, blood banks and genetic disease. In addition, it has great potential for use in field tests and doctors' offices due to its simple and isothermal amplification format.

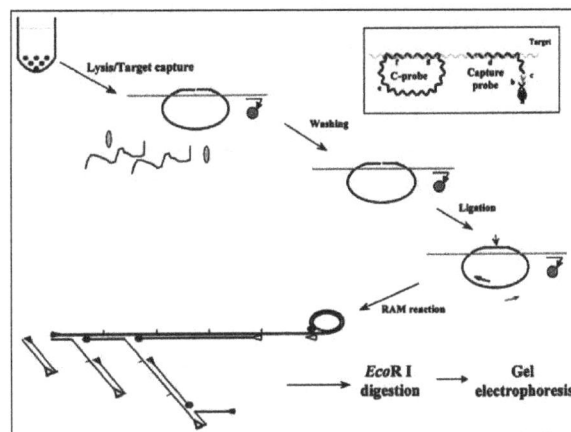

Figure: Schematic representation of ramification amplification of ligated circular probe

Gene Expression

The central dogma of biology describes the method by which information is taken from genes and used to create proteins. DNA transcription produces RNA, then RNA translation makes proteins. This process is known as gene expression and all life forms use it to create the building blocks of life from genetic information.

A cell expresses only a selection of the genes it contains at any one time, which means that the cell can interpret its genetic code in different ways. Controlling which genes are expressed enables the cell to control its size, shape and functions. The ways in which an organism's cells express the genes they contain affects the organism's phenotype, e.g. which color hair a mouse has, or whether it has hair at all.

What is Gene Expression Profiling and who uses it?

Gene expression profiling measures which genes are being expressed in a cell at any given moment. This method can measure thousands of genes at a time; some experiments can measure the entire genome at once. Gene expression profiling measures mRNA levels, showing the pattern of genes expressed by a cell at the transcription level. This often means measuring relative mRNA amounts in two or more experimental conditions, then assessing which conditions resulted in specific genes being expressed.

Gene expression profiling is used by a variety of biomedical researchers, from molecular biologists to environmental toxicologists. This technology can provide accurate information on gene expression, towards countless experimental goals.

Different techniques are used to determine gene expression. These include DNA microarrays and sequencing technologies. The former measures the activity of specific genes of interest and the latter enables researchers to determine all active genes in a cell.

Once a genome has been sequenced, we know what potential a cell has—what characteristics and function it might have—based on the genes it contains. However, sequencing the genome does not tell us which genes a cell is expressing, or the functions or processes it is carrying out at any given moment. To determine these, we need to work out its gene expression profile. If a gene is being used to make mRNA, it is considered 'on'; if it is not being used to make mRNA, it is considered 'off'.

A gene expression profile tells us how a cell is functioning at a specific time. This is because cell gene expression is influenced by external and internal stimuli, including whether the cell is dividing, what factors are present in the cell's environment, the signals it is receiving from other cells, and even the time of day.

Why use Gene Expression Profiling?

Gene expression profiling enables you to investigate the effects of different conditions on gene expression by altering the environment to which the cell is exposed, and determining which genes are expressed. Alternatively, if you already know a gene is involved in a certain cell behavior, gene expression profiling helps you to determine whether a cell is carrying out this function. For example, certain genes are known to be involved in cell division; if these genes are active in a cell, you can tell the cell is undergoing division, or whether a cell is differentiated.

Gene expression profiling is often used in hypothesis generation. If very little is known about when and why a gene will be expressed, expression profiling under different conditions can help design a hypothesis to test in future experiments. For example, if gene A is expressed only when the cell is exposed to other cells, this gene may be involved in intercellular communication. Further experiments could determine whether this is the case.

Gene profiling can also investigate the effect of drug-like molecules on cellular response. You could identify the gene markers of drug metabolism, or determine whether cells express genes known to be involved in response to toxic environments when exposed to the drug.

Gene profiling can also be used as a diagnostic tool. If cancerous cells express higher levels of certain genes, and these genes code for a protein receptor, this receptor may be involved in the cancer, and targeting it with a drug might treat the disease. Gene expression profiling might then be a key diagnostic tool for people with this cancer.

RNA Expression Analysis by RNA Sequencing

RNA expression patterns are key to predicting and classifying human disease based on specific biomarkers. To understand cellular responses, we must determine how gene expression changes are affected in relation to external stimuli, different environmental conditions and genetic lesions.

Transcriptome sequencing, using next-generation sequencing, lets us discover differentially expressed genes without requiring knowledge of which genes are involved.

Protein-coding RNAs are an important source of information, though non-coding RNA is also significant. Next-generation RNA sequencing enables such analysis, along with:

- 'Digital counting' of RNA molecules for highly quantitative and precise measurements.

- Dynamic ranges to fully capture relevant biological changes.

- The discovery of unknown RNAs (novel transcripts, splice variants, and gene fusions).

- The capture of all RNA types (poly-A+, long non-coding RNA and gene fusions) in a single assay.

- Opportunities to focus, going from complete transcriptome analysis to a handful of pre-selected RNA sequences to balance experimental cost and ease of analysis with discovery potentia.

Gene Expression analysis by Real Time Quantitative PCR

Quantification of mRNA using qPCR can be done using Applied Biosystems™ TaqMan® probe–based analysis and Applied Biosystems™ SYBR™ Green dye–based analysis plus using digital PCR, as discussed subsequently.

qPCR is the gold-standard technique for validating differential gene expression profiles, and enables:

- Quantitation of gene products.

- Microarray validation.

- Pathway analysis.

- Studies of developmental biology.

- Quality control and assay validation.

- siRNA/RNAi experiments.

- Low-fold copy number discrimination (down to two-fold).

- Mid- to high-throughput profiling using the OpenArray platform.

Gene Expression Quantification by Digital PCR

Although qPCR is useful for detecting gene expression changes of two-fold or more, a different approach is needed for measuring less than two-fold changes. Digital PCR (dPCR) can be used to resolve low-fold gene expression changes. dPCR enables:

- Absolute quantification of nucleic acid standards and next-generation sequencing libraries.

- Rare target detection.

- Enrichment and separation of mixtures.

Gene Expression Analysis

Gene expression analysis is most simply described as the study of the way genes are transcribed to synthesize functional gene products — functional RNA species or protein products. The study of gene regulation provides insights into normal cellular processes, such as differentiation, and abnormal or pathological processes.

Mechanisms of Gene Regulation

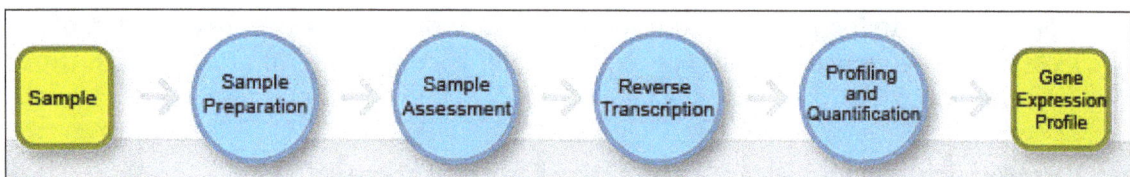

Gene Expression Workflow.

Researchers may perform gene expression analysis at any one of several different levels at which gene expression is regulated: transcriptional, post-transcriptional, translational, and post-translational protein modification.

Transcription, the process of creating a complementary RNA copy of a DNA sequence, can be regulated in a variety of ways. Transcriptional regulation processes are the most commonly studied and manipulated in typical gene expression analysis experiments.

The binding of regulatory proteins to DNA binding sites is the most direct method by which transcription is naturally modulated. Alternatively, regulatory processes can also interact with the transcriptional machinery of a cell. More recently, the influence of epigenetic regulation, such as the effect of variable DNA methylation on gene expression, has been uncovered as a powerful tool for gene expression profiling. Varying degrees of methylation are known to affect chromatin folding and strongly affect accessibility of genes to active transcription.

Following transcription, eukaryotic RNA is typically spliced to remove noncoding intron sequences and capped with a poly(A) tail. At this post-transcriptional level, RNA stability has a significant effect on functional gene expression, that is, the production of functional protein. Small interfering RNA (siRNA) consists of double-stranded nucleic acid molecules that are participants in the RNA interference pathway, in which the expression of specific genes is modulated (typically by decreasing activity). Precisely how this modulation is accomplished is not yet fully understood. A growing field of gene expression analysis is in the area of microRNAs (miRNAs), short RNA molecules that also act as eukaryotic post-transcriptional regulators and gene silencing agents.

Gene Expression Profiling and Quantitation: Methods and Techniques

Researchers studying gene expression employ a wide variety of molecular biology techniques and experimental methods. Gene expression analysis studies can be broadly divided into four areas: RNA expression, promoter analysis, protein expression, and post-translational modification.

RNA Expression

- Northern blotting — steady-state levels of mRNA are directly quantitated by electrophoresis and transfer to a membrane followed by incubation with specific probes. The RNA-probe complexes can be detected using a variety of different chemistries or radionuclide labeling. This relatively laborious technique was the first tool used to measure RNA levels.

- DNA microarrays — an array of oligonucleotide probes bound to a chip surface enables gene expression profiling of many genes in response to a condition. Labeled cDNA from a sample is hybridized to complementary probe sequences on the chip, and strongly associated complexes are identified optically. Gene expression profiling is often a first step in a gene expression analysis workflow, investigating changes in the expression profile of a whole system or examining the effects of mutations in biological systems.

- Real-Time PCR — steady-state levels of mRNA are quantitated by reverse transcription of the RNA to cDNA followed by quantitative PCR (qPCR) on the cDNA. The amount of each specific target is determined by measuring the increase in fluorescence signal from DNA-binding dyes or probes during successive rounds of enzyme-mediated amplification. This precise, versatile tool is used to investigate mutations (including insertions, deletions, and single-nucleotide polymorphisms (SNPs)), identify DNA modifications (such as methylation), confirm results from northern blotting or microarrays, and conduct gene expression profiling. Expression levels can be measured relative to other genes (relative quantification) or against a standard (absolute quantification). Real-time PCR is the gold standard in nucleic acid quantification because of its accuracy and sensitivity. Real-time PCR can be used to quantitate mRNA or miRNA expression following conversion to cDNA or to quantitate genomic DNA directly to investigate transcriptional activity.

Promoter Analysis

- Expression of reporter genes/promoter fusions in host cells — promoter activity (transcription rate) is measured in vivo by introducing fusions of various promoter sequences with a gene encoding a product that can be readily measured to monitor activity levels.

- In vitro transcription (nuclear run-on assays) — transcription rates are measured by incubating isolated cell nuclei with labeled nucleotides, hybridizing the resultant product to a membrane (slot blot), and then exposing this to film or other imaging media.

- Gel shift assays — also called electrophoretic mobility shift assays, these are used to study protein-DNA or protein-RNA interactions. DNA or RNA fragments that are tightly associated with proteins (such as transcription factors) migrate more slowly in an agarose or polyacrylamide gel (showing a positional shift). Identifying the associated sequences provides insight into gene regulation.

- Chromatin immunoprecipitation (ChIP) — protein-binding regions of DNA can be identified in vivo. In living cells, DNA and protein are chemically cross-linked, and the resulting complex is precipitated by antibody-coated beads (immunoprecipitation). Following protein digestion and DNA purification, the sequences of the precipitated DNA are determined.

Protein Expression

- Western blotting — quantification of relative expression levels for specific proteins is accomplished by electrophoretically separating extracted cell proteins, transferring them to a membrane, and then probing the bound proteins with antibodies (targeted to antigens of interest) that are subsequently detected using various chemistries or radiolabelling.

- 2-D Gel Electrophoresis — protein expression profiling is achieved by separating a complex mixture of proteins in two dimensions and then staining to detect differences at the whole-proteome level.

- Immunoassays — proteins are quantitated in solution using antibodies that are bound to color-coded beads (as in the Bio-Plex supension array system) or immobilized to a surface (ELISA), which is subsequently probed with an antibody suspension and is typically detected using a chromogenic or fluorogenic reporter.

Posttranslational Modification Analysis

- Immunoassays — levels of protein phosphorylation and other post-translational modifications are detected using antibodies that are specific for these adducts.

- Mass spectrometry — proteins and their modifications are identified based on their mass.

Serial Analysis of Gene Expression

Serial analysis of gene expression (SAGE) uses mRNA from a particular sample to create complementary DNA (cDNA) fragments which are then amplified and sequenced using high-throughput sequencing technology.

The mechanism behind SAGE is based on tags which can identify the original transcript, and rapid sequencing of chains of tags linked together. The procedure essentially simplifies sequencing by linking the cDNA segments together in a long chain.

The resulting analysis gives a snapshot of the transcriptome of the sample, including the identity and abundance of each mRNA.

Steps of SAGE

SAGE is a complex protocol with many steps:

- Step 1: mRNA is isolated from the sample and reverse transcribed using biotinylated primers to generate cDNA.

- Step 2: cDNA is bound via biotin to streptavidin microbeads.

- Step 3: cDNA is cleaved with restriction enzymes freeing it from the beads.

- Step 4: Cleaved DNA is washed out, leaving truncated cDNA bound to the beads.

- Step 5: Two oligonucleotides with sticky ends are added to the remaining truncated cDNA, in separate samples.

- Step 6: Cleaved DNA is "tagged" enzymatically, removing it from the beads.

- Step 7: Sticky ends are repaired with DNA polymerase.

- Step 8: Blunt ended tags from the two separate samples are ligated together, generating ditags with two different oligonucleotide adapter ends.

- Step 9: Ditags are cleaved to remove the oligonucleotides. Ditags will form long cDNA chains, or concatemers.

- Step 10: Transform concatemers into bacteria for replication.

- Step 11: Isolate concatemers from bacteria and sequence.

Challenges when using SAGE

One challenge is that the tags are only about 13 or 14 base pairs. It can be difficult to identify such a short tag if it's from an unknown gene.

The flip side of that problem is that SAGE can be used to find unknown genes, and in some studies it's an advantage to be able to measure gene expression quantitatively without prior sequence information.

Tags may also have issues with specificity; multiple genes could share the same tag if there is an overlap in sequence. There also can be inconsistencies with the restriction enzymes, and incompatibilities for certain species.

SAGE and DNA microarray

SAGE is similar in many ways to a DNA microarray; however, in a DNA microarray, the mRNAs hybridize to cDNA probes on the array. In SAGE, the data output is based on sequencing. That means SAGE analysis is more quantitative and it does not depend on the use of known genes.

Microarray experiments are generally less costly, and so are used more often in larger-scale studies.

Application

A study of new markers in cancer illustrates how SAGE can be used in biomedical research. Researchers compared gene expression levels in cancerous tissues with those in non-cancerous tissues to search for markers that could diagnose the pancreatic cancer at an early stage. Because the results of a SAGE analysis of a large number of representative tissues had already been published online, the scientists were able to search the database for genes preferentially expressed in pancreatic cancer. From this, they were able to identify a gene calledprostate stem cell antigen (PCSA), that had previously not been associated with pancreatic cancer.

Whole-Genome Sequencing

Whole-genome sequencing (WGS) is a comprehensive method for analyzing entire genomes. Genomic information has been instrumental in identifying inherited disorders, characterizing the mutations that drive cancer progression, and tracking disease outbreaks. Rapidly dropping sequencing costs and the ability to produce large volumes of data with today's sequencers make whole-genome sequencing a powerful tool for genomics research.

While this method is commonly associated with sequencing human genomes, the scalable, flexible nature of next-generation sequencing (NGS) technology makes it equally useful for sequencing any species, such as agriculturally important livestock, plants, or disease-related microbes.

Advantages of Whole-Genome Sequencing

- Provides a high-resolution, base-by-base view of the genome.
- Captures both large and small variants that might otherwise be missed.
- Identifies potential causative variants for further follow-on studies of gene expression and regulation mechanisms.
- Delivers large volumes of data in a short amount of time to support assembly of novel genomes.

Clone-by-clone

This method requires the genome to have smaller sections copied and inserted into bacteria. The bacteria then can be grown to produce identical copies, or "clones," containing approximately 150,000 base pairs of the genome that is desired to be sequenced. Then, the inserted DNA in each clone is further broken down into smaller, overlapping 500 base pair chunks. These smaller inserts are sequenced. After sequencing is performed, the overlapping portions are used to reassemble the clone. This approach was used to sequence the first human genome using Sanger sequencing. This approach is time-consuming and costly, but it is reliable.

Whole-Genome Shotgun

As the name implies, "shotgun" sequencing is a method that breaks DNA into small random pieces for sequencing and reassembly. The pieces of DNA are also cloned into bacteria for growth, isolation and subsequent sequencing. Because the pieces are random, there are overlapping sequences that aid in reassembly into the original DNA order. This approach was originally used in Sanger sequencing but is now also used in next-generation sequencing methods providing rapid genome sequencing with lower costs. It is only good for shorter "reads" (ie, sequencing on shorter DNA fragments to be put back together again). Because it is reassembled based on overlapping regions and has shorter read lengths, it is best utilized when a reference genome is available, and it requires sophisticated computational approaches to reassemble the sequence. It also can be challenging for genomes with many repetitive regions.

Assembly of Sequencing Reads

Because genomes are sequenced in varying lengths of DNA fragments, the resulting sequences must be put back together. This is referred to as "assembly," or "reassembly." Two common approaches are de novo assembly and assembly by reference mapping.

De novo assembly is performed by identifying overlapping regions in the DNA sequences, aligning the sequences and putting them back together to form the genome. This is done without any sequence with which to compare. Mapping to a reference genome uses another genome to align new sequencing data to as a comparator.

Although de novo assembly can be challenging, this approach is the only one available for sequencing new organisms. Additionally, de novo assembly introduces results with less bias than mapping to a reference genome. Mapping to a reference genome is easier and requires less contiguous reads, but new or unexpected sequences can be lost. The sequence results obtained by this method is only as good as the reference genome chosen; however, it can provide better identification of single nucleotide polymorphisms (SNPs). Multiple institutions and genomic sequencing companies have invested considerable time and effort into creating improved reference genomes. Single nucleotide polymorphisms are known to vary by race and ethnicity, thus, multiple reference genomes have been created for various races/ethnicities.

Examples of Next-generation Sequencing Platforms

Several companies focus on development and marketing of next-generation sequencing machines (often referred to as "platforms") for use in whole-genome (and other) sequencing. Illumina is considered by many as the leader because of the number of users that utilize its systems.

Illumina has multiple platforms depending on the need. The Illumina HiSeq is one of the more common sequencers found in laboratories, including major research institutions, companies providing next-generation sequencing services for clinics and labs, and pathology laboratories. It has a high throughput, capable of sequencing many genomes rapidly with reasonable costs. This instrument also can be used to look at copy number variation, as well as mutations and other alterations, and RNA expression levels to do transcriptomics. Because of the popularity in the clinic of targeted sequencing panels, which are much smaller with clinics requiring faster turnaround times for treatment of patients, Illumina created the MiSeq, which can provide same-day sequencing results for very small panels. Illumina also produced multiple variations to provide sequencers for each disease area optimizing output, turnaround time and costs for specific use cases.

Thermo Fisher Scientific's Ion Torrent or Ion Proton uses a completely different technology based on detection of pH differences and was once expected to provide better utility for clinical applications because it was easier to use, cost less and provided faster turnaround time. However, Illumina countered with new machines to fit these needs. Consequently, both are found in research and clinical laboratories.

Other technologies developed recently use different novel approaches. A few examples are provided below.

Oxford Nanopore Technologies introduced the MinION, which enables anyone to sequence on a desktop computer using a USB device. The DNA is passed through a protein nanopore membrane for sequencing and detection by creation of an ionic current that varies based on the nucleotide.

Pacific Biosciences introduced its single molecule, real-time technology with the longest reads to date, with average read lengths of more than 10,000 base pairs compared with more than 150 base pairs. Single molecule, real-time technology uses a chip with single DNA molecules attached. Zero-mode waveguide technology enables isolation of a single nucleotide for the DNA polymerase to add fluorescent labels for detection of each base. The error rate of this instrument is still higher than some of the prior technologies, but a lot of interest has been generated, and there is hope that speed and costs can be further optimized with the new approach.

Coverage Breadth and Depth

Coverage refers to the number of reads that show a specific nucleotide in the reconstructed DNA sequence. A read is a string of A, T, C, G bases that correspond to the reference DNA. There are millions of reads in a sequencing run. Increased coverage depth results in increased confidence in variant identification.

For the human genome, a 10- to 30-times coverage depth is acceptable for detecting mutations, SNPs and rearrangements. A next-generation sequencing approach that provides a coverage depth of 30 times is considered to have high coverage. However, as coverage depth increases, coverage breadth decreases.

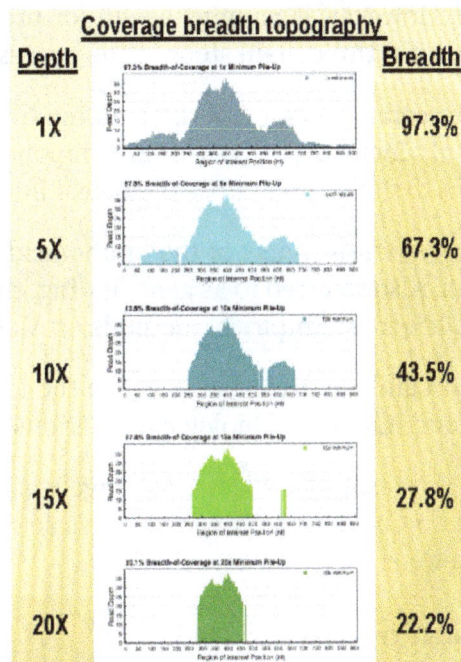

Figure: Relationship between coverage breadth vs. coverage depth.

Gene Annotation

genomes make the basic genetic material and typically consist of DNA. Whereby, genome include the genes (coding) and the non-coding regions, of interest to us, are the coding regions as they actively influence basic life processes. The genes contain useful biological information that is required in building up and maintaining an organism. Gene annotation can be defined merely as the process of making nucleotide sequence meaningful. However, it's a much complex process encompassing several procedures and a broad range of activities.

Gene annotation involves the process of taking the raw DNA sequence produced by the genome-sequencing projects and adding layers of analysis and interpretation necessary to extracting biologically significant information and placing such derived details into context. Through the aid of bioinformatics, there exists software to perform such complex procedures. The first gene annotation software system was developed in 1995 at The Institute for Genomic Research, and this was used to sequence and analyze the genes of the bacterium *Haemophilus influenza*.

As a process of identification of gene location and coding regions, gene annotation helps us have an insight of what these genes do in the body by establishing structural aspects and relating them to functions of different proteins. Currently, the process is automated, and the National Center for Biomedical Ontology have a database for records and to enable comparison.

How is Gene Annotation Performed?

Gene annotation can either be manual or electronic with the aid of tools developed by an amalgamation of organizations. The downsides of the manual technique are that it is time-consuming and the turn-over rate is much low. However, it remains useful for predictive purposes thus serves a complementary function. There exist three main steps in the process of gene annotation:

- Identification of the non-coding regions of the genome (exons). This is vital to limit the range of analysis and only focus on the essential components as it is needless doing the tedious work on portions that give no or little biological information.

- Gene prediction; these give an overview of the amino acid components of the genes and the role of such elements. Also referred to as gene finding, this process identifies regions of genomic DNA that encode genes. Empirical methods or Ab Initio methods can do it.

- Establishing a connection and a correlation between the identified elements and the biological information at hand. Linking of biological functions and data is possible this way.

- Homology-based tools for example Blast has hugely simplified the process of gene annotation, and this can now be done without much hassle as witnessed in manual methods that require human expertise.

Modalities of Gene Annotation

Genomics is a broad study and can be subdivided as structural genomics, functional genomics, and comparative genomics to leverage the understanding of this crucial topic. Similarly, gene annotation exists as a double-phased entity comprising of structural gene annotation and functional gene annotation.

Structural Annotation

The initial process in gene annotation and involve identification by physical appearance, chemical composition, molecular weight variations, and general morphology. Such differences as coding regions, gene structures, ORFs and their locations, as well as regulatory motifs, are crucial information that is derived from this procedure and influence the process of gene identification as well as distinction. The accuracy of this process can be evaluated based on two parameters; specificity and accuracy. Where sensitivity is the percentage of right signals predicted among all possible correct strengths while specificity refers to the proportion of right signal among all that are forecasted.

Functional Annotation

The process of relating crucial biological functions to the genetic elements as depicted in the structural annotation step. Biochemical functions, physiological functions, involved regulations and interactions atop expressions are some of the critical roles that are often considered in DNA annotation.

The above steps can involve biological experiments as well as in silico analysis mimicking the internal conditions. A new method seeking to improve genomics annotation-Proteogenomics is currently in use, and it utilizes information from expressed proteins, such information is obtained from mass spectrometry.

Essential Components

Gene annotation is a purposeful process, and some of the vital information that we seek to extract from this process include; CDs, mRNA, Pseudogenes, promoter and poly-A signals, mcRNA among others. Such elements are minute and identification may be hectic. Scientists have developed software and tools to aid the process and notable tools frequently used are; ORF detectors, promoter detectors and start/stop codon identifiers. Automation of this process has created enhanced accuracy, and now there exist large discrepancies between with the manually conducted procedures as gene sequencing is a dynamic topic.

Transcriptome Sequencing

Transcriptome sequencing encompasses a wide variety of applications from simple mRNA profiling to discovery and analysis of the entire transcriptome, including both coding mRNA and non-coding RNA (e.g., miRNA, small RNAs, linc RNAs). These applications, collectively called RNA-Seq, are extremely popular for next generation sequencing platforms as they uncover information that may be missed by array-based platforms, as no prior knowledge of the transcript sequence is needed. RNA-Seq was initially used primarily for discovery applications (rare genes, splice junctions, gene fusions) and with novel or poorly studied organisms for which there weren't good standard microarrays. However, as the technology is becoming more available and costs are coming down, it is starting to be used for RNA profiling (sample comparison) as well. Also, as it is sequencing based, it is well suited for specialty applications such as RNA editing and allele specific expression. While there are a variety of RNA-Seq applications and protocols, most follow the basic strategy of isolating RNA (such as with poly dT to pull down mRNA), converting it to DNA and then adding adaptor sequences to generate a library suitable for sequencing.

mRNA-Seq

One of the most popular forms of transcriptome sequencing is mRNA-Seq, which targets all polyadenylated mRNA transcripts, or the coding portion of the transcriptome. The depth offered by next generation sequencing is leveraged to find new genes that were undetectable before due to their low level of expression. By the same token, the increased depth and reduced cost of sequencing means these projects can be used to profile gene expression while differentiating between isoforms of the same gene via paired-end reads.

Small RNA-Seq

Attempts to target small RNA for sequencing have been driven by interest in microRNAs, as they act as regulatory elements in the transcriptome. By complexing in the 3′ region with mRNA transcripts that have complementary sequence, microRNAs mark these transcripts for degradation prior to translation. The depth offered by next generation sequencing again loans itself to both novel microRNA gene discovery and expression profiling. The major problem in preparing these samples is the difficulty in separating these species from similarly sized adapter-dimers. Gel-based size selections that recover adapter-bound small RNAs while excluding adapter-dimers from the sample are employed here.

Tag-based Approaches

The original RNA sequencing method used with next generation sequencing was 'digital gene expression' (DGE), a modified form of 'serial analysis of gene expression' (SAGE). In this method a single short tag (~23 bases) is generated for each transcript as defined by the relation of a four base restriction enzyme recognition site and the location of the polyA tail. (In practice, additional tags can be generated by the presence of polyA stretches elsewhere in the transcript.) Because each transcript contains only a single tag (or at the most, a few), DGE libraries are substantially less complex than standard RNA-Seq libraries, allowing useful results with fewer reads per sample (1-2M compared with ~30M). Alternatively, higher read depths allow for the detection of extremely rare transcripts, which can be useful when looking for single or low copy transcripts in a population of cells. The SuperSAGE method improves on the original DGE methods through the generation of larger tags, allowing for more precise alignment to the transcriptome. While these methods still have adherents, the rapidly falling costs of sequencing are diminishing the importance of needing fewer reads, while the more comprehensive coverage of the transcriptome by other RNA-Seq methods is gaining in popularity. However, variants of these "tag profiling" methods, such as 'cap analysis gene expression' (CAGE), can be very useful for determining the precise 5' ends of transcripts.

Whole Transcriptome Sequencing

A truly comprehensive view of the transcriptome requires the examination of all unique RNA transcripts for an organism, including both the coding and non-coding portions. This is achieved by combining the results of various RNA-Seq methods. The primary method is the generation of RNA-Seq libraries without the use of polyA enrichment, allowing for the sequencing of a significant portion of the non-coding RNA. However, since rRNA accounts for 95-98% of the transcriptome, it is still necessary to eliminate this fraction prior to sequencing. This is often accomplished with commercially available kits that can use oligos complementary to known rRNA sequences to sequester those transcripts. Alternatively, 'duplex-specific nuclease' (or DSN) can be used to eliminate highly abundant sequences, greatly enriching the other portions of the transcriptome. The standard RNA-Seq methods generally don't capture small RNAs well, so the results are combined with small RNA-Seq and possibly one or more of the CAGE methods to more accurately define the transcription start and stop sites.

RNA Sequencing and Analysis

RNA sequencing (RNA-Seq) uses the capabilities of high-throughput sequencing methods to provide insight into the transcriptome of a cell. Compared to previous Sanger sequencing- and microarray-based methods, RNA-Seq provides far higher coverage and greater resolution of the dynamic nature of the transcriptome. Beyond quantifying gene expression, the data generated by RNA-Seq facilitate the discovery of novel transcripts, identification of alternatively spliced genes, and detection of allele-specific expression. Recent advances in the RNA-Seq workflow, from sample preparation to library construction to data analysis, have enabled researchers to further elucidate the functional complexity of the transcription. In addition to polyadenylated messenger RNA (mRNA) transcripts, RNA-Seq can be applied to investigate different populations of RNA, including total RNA, pre-mRNA, and noncoding RNA, such as microRNA and long ncRNA.

The central dogma of molecular biology outlines the flow of information that is stored in genes as DNA, transcribed into RNA, and finally translated into proteins. The ultimate expression of this genetic information modified by environmental factors characterizes the phenotype of an organism. The transcription of a subset of genes into complementary RNA molecules specifies a cell's identity and regulates the biological activities within the cell. Collectively defined as the transcriptome, these RNA molecules are essential for interpreting the functional elements of the genome and understanding development and disease.

The transcriptome has a high degree of complexity and encompasses multiple types of coding and noncoding RNA species. Historically, RNA molecules were relegated as a simple intermediate between genes and proteins, as encapsulated in the central dogma of molecular biology. Therefore, messenger RNA (mRNA) molecules were the most frequently studied RNA species because they encoded proteins via the genetic code. In addition to protein-coding mRNA, there is a diverse group of noncoding RNA (ncRNA) molecules that are functional. Previously, most known ncRNAs fulfilled basic cellular functions, such as ribosomal RNAs and transfer RNAs involved in mRNA translation, small nuclear RNA (snRNAs) involved in splicing, and small nucleolar RNAs (snoRNAs) involved in the modification of rRNAs. More recently, novel classes of RNA have been discovered, enhancing the repertoire of ncRNAs. For instance, one such class of ncRNAs is small noncoding RNAs, which include microRNA (miRNA) and piwi-interacting RNA (piRNA), both of which regulate gene expression at the posttranscriptional level. Another noteworthy class of ncRNAs is long noncoding RNAs (lncRNAs). As a functional class, lncRNAs were first described in mice during the large-scale sequencing of cDNA libraries. A myriad of molecular functions have been discovered for lncRNAs, including chromatin remodeling, transcriptional control, and post-transcriptional processing, although the vast majority are not fully characterized.

Initial gene expression studies relied on low-throughput methods, such as northern blots and quantitative polymerase chain reaction (qPCR), that are limited to measuring single transcripts. Over the last two decades, methods have evolved to enable genome-wide quantification of gene expression, or better known as transcriptomics. The first transcriptomics studies were performed using hybridization-based microarray technologies, which provide a high-throughput option at relatively low cost. However, these methods have several limitations: the requirement for a priori knowledge of the sequences being interrogated; problematic cross-hybridization artifacts in the analysis of highly similar sequences; and limited ability to accurately quantify lowly expressed and very highly expressed genes. In contrast to hybridization-based methods, sequence-based approaches have been developed to elucidate the transcriptome by directly determining the transcript sequence. Initially, the generation of expressed sequence tag (EST) libraries by Sanger sequencing of complementary DNA (cDNA) was used in gene expression studies, but this approach is relatively low-throughput and not ideal for quantifying transcripts. To overcome these technical constraints, tag-based methods such as serial analysis of gene expression (SAGE) and cap analysis gene expression (CAGE) were developed to enable higher throughput and more precise quantification of expression levels. By quantifying the number of tagged sequences, which directly corresponded to the number of mRNA transcripts, these tag-based methods provide a distinct advantage over measuring analog-style intensities as in array-based methods. However, these assays are insensitive to measuring expression levels of splice isoforms and cannot be used for novel gene discovery. In addition, the laborious cloning of sequence tags, the high cost of automated Sanger sequencing, and the requirement for large amounts of input RNA have greatly limited its use.

The development of high-throughput next-generation sequencing (NGS) has revolutionized transcriptomics by enabling RNA analysis through the sequencing of complementary DNA (cDNA). This method, termed RNA sequencing (RNA-Seq), has distinct advantages over previous approaches and has revolutionized our understanding of the complex and dynamic nature of the transcriptome. RNA-Seq provides a more detailed and quantitative view of gene expression, alternative splicing, and allele-specific expression. Recent advances in the RNA-Seq workflow, from sample preparation to sequencing platforms to bioinformatic data analysis, has enabled deep profiling of the transcriptome and the opportunity to elucidate different physiological and pathological conditions.

Transcriptome Sequencing

The introduction of high-throughput next-generation sequencing (NGS) technologies revolutionized transcriptomics. This technological development eliminated many challenges posed by hybridization-based microarrays and Sanger sequencing-based approaches that were previously used for measuring gene expression. A typical RNA-Seq experiment consists of isolating RNA, converting it to complementary DNA (cDNA), preparing the sequencing library, and sequencing it on an NGS platform. However, many experimental details, dependent on a researcher's objectives, should be considered before performing RNA-Seq. These include the use of biological and technical replicates, depth of sequencing, and desired coverage across the transcriptome. In some cases, these experimental options will have minimal impact on the quality of the data. However, in many cases the researcher must carefully design the experiment, placing a priority on the balance between high-quality results and the time and monetary investment.

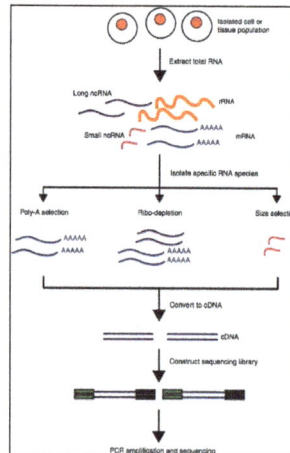

Figure: Overview of RNA-Seq. First, RNA is extracted from the biological material of choice (e.g., cells, tissues).

Second, subsets of RNA molecules are isolated using a specific protocol, such as the poly-A selection protocol to enrich for polyadenylated transcripts or a ribo-depletion protocol to remove ribosomal RNAs. Next, the RNA is converted to complementary DNA (cDNA) by reverse transcription and sequencing adaptors are ligated to the ends of the cDNA fragments. Following amplification by PCR, the RNA-Seq library is ready for sequencing.

Isolation of RNA

The first step in transcriptome sequencing is the isolation of RNA from a biological sample. To ensure a successful RNA-Seq experiment, the RNA should be of sufficient quality to produce a library for sequencing. The quality of RNA is typically measured using an Agilent Bioanalyzer, which produces an RNA Integrity Number (RIN) between 1 and 10 with 10 being the highest quality samples showing the least degradation. The RIN estimates sample integrity using gel electrophoresis and analysis of the ratios of 28S to 18S ribosomal bands. Note that the RIN measures are based on mammalian organisms and certain species with abnormal ribosomal ratios (i.e., insects) may erroneously generate poor RIN numbers. Low-quality RNA (RIN < 6) can substantially affect the sequencing results (e.g., uneven gene coverage, 3′–5′ transcript bias, etc.) and lead to erroneous biological conclusions. Therefore, high-quality RNA is essential for successful RNA-Seq experiments. Unfortunately, high-quality RNA samples may not be available in some cases, such as human autopsy samples or paraffin embedded tissues, and the effect of degraded RNA on the sequencing results should be carefully considered.

Library Preparation Methods

Following RNA isolation, the next step in transcriptome sequencing is the creation of an RNA-Seq library, which can vary by the selection of RNA species and between NGS platforms. The construction of sequencing libraries principally involves isolating the desired RNA molecules, reverse-transcribing the RNA to cDNA, fragmenting or amplifying randomly primed cDNA molecules, and ligating sequencing adaptors. Within these basic steps, there are several choices in library construction and experimental design that must be carefully made depending on the specific needs of the researcher. Additionally, the accuracy of detection for specific types of RNAs is largely dependent on the nature of the library construction. Although there are a few basic steps for preparing RNA-Seq libraries, each stage can be manipulated to enhance the detection of certain transcripts while limiting the ability to detect other transcripts.

Table: RNA-Seq library protocols

Library design	Usage	Description
Poly-A selection	Sequencing mRNA	Select for RNA species with poly-A tail and enriches for mRNA
Ribo-depletion	Sequencing mRNA, pre-mRNA, ncRNA	Removes ribosomal RNA and enriches for mRNA, pre-mRNA, and ncRNA
Size selection	Sequencing miRNA	Selects RNA species using size fractionation by gel electrophoresis
Duplex-specific nuclease	Reduce highly abundant transcripts	Cleaves highly abundant transcripts, including rRNA and other highly expressed genes
Strand-specific	De novo transcriptome assembly	Preserves strand information of the transcript
Multiplexed	Sequencing multiple samples together	Genetic barcoding method that enables sequencing multiple samples together
Short-read	Higher coverage	Produces 50–100 bp reads; generally higher read coverage and reduced error rate compared to long-read sequencing
Long-read	De novo transcriptome assembly	Produces >1000 bp reads; advantageous for resolving splice junctions and repetitive regions

Selection of RNA Species

Before constructing RNA-Seq libraries, one must choose an appropriate library preparation protocol that will enrich or deplete a "total" RNA sample for particular RNA species. The total RNA pool includes ribosomal RNA (rRNA), precursor messenger RNA (pre-mRNA), mRNA, and various classes of noncoding RNA (ncRNA). In most cell types, the majority of RNA molecules are rRNA, typically accounting for over 95% of the total cellular RNA. If the rRNA transcripts are not removed before library construction, they will consume the bulk of the sequencing reads, reducing the overall depth of sequence coverage and thus limiting the detection of other less-abundant RNAs. Because the efficient removal of rRNA is critical for successful transcriptome profiling, many protocols focus on enriching for mRNA molecules before library construction by selecting for polyadenylated (poly-A) RNAs. In this approach, the 3' poly-A tail of mRNA molecules is targeted using poly-T oligos that are covalently attached to a given substrate (e.g., magnetic beads). Alternatively, researchers can selectively deplete rRNA using commercially available kits, such as RiboMinus (Life Technologies) or RiboZero (Epicentre). This latter method facilitates the accurate quantification of noncoding RNA species, which may be polyadenylated and thus excluded from poly-A libraries. Lastly, highly abundant RNA can be removed by denaturing and re-annealing double-stranded cDNA in the presence of duplex-specific nucleases that preferentially digest the most abundant species, which re-anneal as double-stranded molecules more rapidly than less-abundant molecules. This method can also be used to remove other highly abundant mRNA transcripts in samples, such as hemoglobin in whole blood, immunoglobulins in mature B cells, and insulin in pancreatic beta cells.

A comprehensive understanding of the technical biases and limitations surrounding each methodological approach is essential for selecting the best method for library preparation. For example, poly-A libraries are the superior choice if one is solely interested in coding RNA molecules. Conversely, ribo-depletion libraries are a more appropriate choice for accurately quantifying noncoding RNA as well as pre-mRNA that has not been posttranscriptionally modified. Furthermore, moderate differences exist between ribo-depletion protocols, such as the efficiency of rRNA removal and differential coverage of small genes, which should be investigated before selecting a method.

In addition to the selective depletion of specific RNA species, new approaches have been developed to selectively enrich for regions of interest. These approaches include methods employing PCR-based approaches, hybrid capture, in-solution capture, and molecular inversion probes. The hybridization-based in solution capture involves a set of biotinylated RNA baits transcribed from DNA template oligo libraries that contain sequences corresponding to particular genes of interest. The RNA baits are combined with the RNA-Seq library where they hybridize to RNA sequences that are complementary to the baits, and the bounded complexes are recovered using streptavidin-coated beads. The resulting RNA-Seq library is now enriched for sequences corresponding to the baits and yet retains its gene expression information despite the removal of other RNA species. The approach enables researchers to reduce sequencing costs by sequencing selected regions in a greater number of samples.

Selection of Small RNA Species

Complementing the library preparation protocols discussed above, more specific protocols have been developed to selectively target small RNA species, which are key regulators of gene expression. Small RNA species include microRNA (miRNA), small interfering RNA (siRNA), and piwi-interacting RNA (piRNA). Because small RNAs are lowly abundant, short in length (15–30 nt), and lack polyadenylation, a separate strategy is often preferred to profile these RNA species. Similar to total RNA isolation, commercially available extraction kits have been developed to isolate small

RNA species. Most kits involve isolation of small RNAs by size fractionation using gel electrophoresis. Size fractionation of small RNAs requires involves running the total RNA on a gel, cutting a gel slice in the 14–30 nucleotide region, and purifying the gel slice. For higher concentrations of small RNAs, the excised gel slice can be concentrated by ethanol precipitation. An alternative to gel electrophoresis is the use of silica spin columns, which bind and elute small RNAs from a silica column. After isolation of small RNAs species from total RNA, the RNA is ready for cDNA synthesis and primer ligation.

cDNA Synthesis

Universal to all RNA-Seq preparation methods is the conversion of RNA into cDNA because most sequencing technologies require DNA libraries. Most protocols for cDNA synthesis create libraries that were uniformly derived from each cDNA strand, thus representing the parent mRNA strand and its complement. In this conventional approach, the strand orientation of the original RNA is lost as the sequencing reads derived from each cDNA strand are indistinguishable in an effort to maximize efficiency of reverse transcription. However, strand information can be particularly valuable for distinguishing overlapping transcripts on opposite strands, which is critical for de novo transcript discovery. Therefore, alternative library preparation protocols have since been developed that yield strand-specific reads. One strategy to preserve strand information is to ligate adapters in predetermined directions to single-stranded RNA or the first-strand of cDNA. Unfortunately, this approach is laborious and results in coverage bias at both the 5′ and 3′ ends of cDNA molecules. The preferred strategy to preserve strandedness is to incorporate a chemical label such as deoxy-UTP (dUTP) during synthesis of the second-strand cDNA that can be specifically removed by enzymatic digestion. During library construction, this facilitates distinguishing the second-strand cDNA from the first strand. Although this approach is favored, the validity of antisense transcripts near highly expressed genes should be measured with caution because a small amount of reads (~1%) have been observed from the opposite strand.

Multiplexing

Another consideration for constructing cost-effective RNA-Seq libraries is assaying multiple indexed samples in a single sequencing lane. The large number of reads that can be generated per sequencing run (e.g., a single lane of an Illumina HiSeq 2500 generates up to 750 million paired-end reads) permits the analysis of increasingly complex samples. However, increasingly high sequencing depths provide diminishing returns for lower complexity samples, resulting in oversampling with minimal improvement in data quality. Therefore, an affordable and efficient solution is to introduce unique 6-bp indices, also known as "barcodes," to each RNA-Seq library. This enables the pooling and sequencing of multiple samples in the same sequencing reaction because the barcodes identify which sample the read originated from. Depending on the application, adequate transcriptome coverage can be attained for 2–20 samples. To detect transcripts of moderate to high abundance, ~30–40 million reads are required to accurately quantify gene expression. To obtain coverage over the full-sequence diversity of complex transcript libraries, including rare and lowly-expressed transcripts, up to 500 million reads is required. As such, for any given study it is important to consider the level of sequencing depth required to answer experimental questions with confidence while efficiently using NGS resources.

Quantitative Standards

Although RNA-Seq is a widely used technique for transcriptome profiling, the rapid development of sequencing technologies and methods raises questions about the performance of different platforms and protocols. Variation in RNA-Seq data can be attributed to an assortment of factors, ranging from the NGS platform used to the quality of input RNA to the individual performing the experiment. To control for these sources of technical variability, many laboratories use positive controls or "spike-ins" for sequencing libraries. The External RNA Controls Consortium (ERCC) developed a set of universal RNA synthetic spike-in standards for microarray and RNA-Seq experiments. The spike-ins consist of a set of 96 DNA plasmids with 273–2022 bp standard sequences inserted into a vector of ~2800 bp. The spike-in standard sequences are added to sequencing libraries at different concentrations to assess coverage, quantification, and sensitivity. These RNA standards serve as an effective quality control tool for separating technical variability from biological variability detected in differential transcriptome profiling studies.

Selection of Tissue or Cell Populations

When beginning an RNA-Seq experiment, one of the initial considerations is the choice of biological material to be used for library construction and sequencing. This choice is not trivial considering there are hundreds of cell types in over 200 different tissues that make up greater than 50 unique organs in humans alone. In addition to spatial (e.g., cell- and tissue-type) specificity, gene expression shows temporal specificity, such that different developmental stages will show unique expression signatures. Ultimately, the biological material chosen will be dependent on both the experimental goals and feasibility. For example, the tissue of choice for an investigation of unique gene expression signatures in colon cancer, the tissue choice is clear. However, for research studies investigating variation in gene expression across individuals in a population, the choice of biological material is less apparent and will likely depend on the feasibility of obtaining the biological samples (e.g., blood draws are less invasive and easier to perform than tissue biopsies).

Handling Tissue Heterogeneity

Another consideration when selecting the biological source of RNA is the heterogeneity of tissues. The accuracy of gene expression quantification is dependent on the purity of samples. In fact, the heterogeneity can substantially impact estimations of transcript abundances in samples composed of multiple cell types. Most tissue samples isolated from the human body are heterogeneous by nature. Furthermore, pathological tissue samples are often composed of disease-state cells surrounded by normal cells. To isolate distinct cell types, experimental methods have been developed, including laser-capture microdissection and cell purification. Laser-capture microdissection enables the isolation of cell types that are morphologically distinguishable under direct microscopic visualization. Although this technique yields high-quality RNA, the total yield is low and requires PCR amplification, thereby introducing amplification biases and creating less distinguishable expression profiles across different cell types. Cell purification and enrichment protocols are also available, such as differential centrifugation and fluorescence-activated cell sorting. In conjunction with RNA-Seq, these experimental methods have overcome previous technical limitations and enable researchers to uncover unique expression signatures across specific cell-types and developmental stages. In addition to these experimental methods, in silico probabilistic models can be applied in downstream analysis to differentiate the transcript abundances of distinct cells from RNA-Seq data of heterogeneous tissue samples. Interestingly, in some cases, the sample heterogeneity can have advantages in transcriptome profiling by identifying novel pathways, implicating cellular origins of disease, or identifying previously unknown pathological sites.

Single-Cell Transcriptomics

Beyond tissue heterogeneity, considerable evidence indicates that cell-to-cell variability in gene expression is ubiquitous, even within phenotypically homogeneous cell populations. Unfortunately, conventional RNA-Seq studies do not capture the transcriptomic composition of individual cells. The transcriptome of a single cell is highly dynamic, reflecting its functionality and responses to ever-changing stimuli. In addition to cellular heterogeneity resulting from regulation, individual cells show transcriptional "noise" that arises from the kinetics of mRNA synthesis and decay. Furthermore, genes that show mutually exclusive expression in individual cells may be observed as genes showing co-expression in expression analyses of bulk cell populations.

To uncover cell-to-cell variation within populations, significant efforts have been invested in developing single-cell RNA-Seq methods. The biggest challenge has been extending the limits of library preparation to accommodate extremely low input RNA. A human cell contains <1 pg of mRNA, whereas most sequencing protocols such as Illumina's TruSeq RNA-Seq kit recommends 400 ng to 1 µg of input RNA material. Various single-cell RNA amplification methods have been developed to accommodate less input RNA. The key limiting factors in the detection of transcripts in single cells are cDNA synthesis and PCR amplification. The efficiency of RNA-to-cDNA conversion is imperfect, estimated to be as low as 5%–25% of all transcripts. In addition, PCR amplification methods do not linearly amplify transcript and are prone to introduce biases based on the nucleic acid composition of different transcripts, ultimately altering the relative abundance of these transcripts in the sequencing library. Methods that avoid PCR amplification steps, such as CEL-Seq, through linear in vitro amplification of the transcriptome can avoid these biases. In addition, the use of nanoliter-scale reaction volumes with microfluidic devices as opposed to microliter-scale reactions can reduce biases that arise during sample preparation. Although single-cell methods are still under active development, quantitative assessments of these techniques indicate that obtaining accurate transcriptome measurements by single-cell RNA-Seq is possible after accounting for technical noise. These methods will undoubtedly be important for uncovering oscillatory and heterogeneous gene expression within single-cell types, as well as identifying cell-specific biomarkers that further our understanding of biology across many physiological and pathological conditions.

Sequencing Platforms for Transcriptomics

When designing an RNA-Seq experiment, the selection of a sequencing platform is important and dependent on the experimental goals. Currently, several NGS platforms are commercially available and other platforms are under active technological development. The majority of high-throughput sequencing platforms use a sequencing-by-synthesis method to sequence tens of millions of sequence clusters in parallel. The NGS platforms can often be categorized as either ensemble-based (i.e. sequencing many identical copies of a DNA molecule) or single-molecule-based (i.e. sequencing a single DNA molecule). The differences between these sequencing techniques and platforms can affect downstream analysis and interpretation of the sequencing data.

In recent years, the sequencing industry has been dominated by Illumina, which applies an ensemble-based sequencing-by-synthesis approach. Using fluorescently labeled reversible-terminator nucleotides, DNA molecules are clonally amplified while immobilized on the surface of a glass flowcell. Because molecules are clonally amplified, this approach provides the relative RNA expression levels of genes. To remove potential PCR-amplification biases, PCR controls and specific steps in the downstream computational analysis are required. One major benefit of ensemble-based platforms is low sequencing error rates (<1%) dominated by single mismatches. Low error rates are particularly important for sequencing miRNAs, whose relatively small sizes result in misalignment or loss of reads if error rates are too high. Currently, the Illumina HiSeq platform is the most commonly applied next-generation sequencing technology for RNA-Seq and has set the standard for NGS sequencing. The platform has two flow cells, each providing eight separate lanes for sequencing reactions to occur. The sequencing reactions can take between 1.5 and 12 d to complete, depending on the total read length of the library. Even more recently, Illumina released the MiSeq, a desktop sequencer with lower throughput but faster turnaround (generates ~30 million paired-end reads in 24 h). The simplified workflow of the MiSeq instrument offers rapid turnaround time for transcriptome sequencing on a smaller scale.

Single-molecule-based platforms such as PacBio enable single-molecule real-time (SMRT) sequencing. This approach uses DNA polymerase to perform uninterrupted template-directed synthesis using fluorescently labeled nucleosides. As each base is enzymatically incorporated into a growing DNA strand, a distinctive pulse of fluorescence is detected in real-time by zero-mode waveguide nanostructure arrays. An advantage of SMRT is that it does not include a PCR amplification step, thereby avoiding amplification bias and improving uniform coverage across the transcriptome. Another advantage of this sequencing approach is the ability to produce extraordinarily long reads with average lengths of 4200 to 8500 bp, which greatly improves the detection of novel transcript structures. A critical disadvantage of SMRT is a high rate of errors (~5%) that are predominately characterized by insertions and deletions; the high error rate results in misalignment and loss of sequencing reads due to the difficulty of matching erroneous reads to the reference genome.

Another important consideration for choosing a sequencing platform is transcriptome assembly. Transcriptome assembly is necessary to transform a collection of short sequencing reads into a set of full-length transcripts. In general, longer sequencing reads make it simpler to accurately and unambiguously assemble transcripts, as well as identify splicing isoforms. The extremely long reads generated by the PacBio platform are ideal for de novo transcriptome assembly in which the reads are not aligned to a reference transcriptome. The longer reads will facilitate an accurate detection of alternative splice isoforms, which may not be discovered with shorter reads. Moleculo, a company acquired by Illumina, has developed long-read sequencing technology capable of producing 8500 bp reads. Although it has yet to be widely adopted for transcriptome sequencing, the long reads aid transcriptome assembly. Lastly, Illumina has developed protocols for its desktops MiSeq to sequence slightly longer reads (up to 350 bp). Although much shorter than PacBio and Moleculo reads, the longer MiSeq reads can also be used to improve both de novo and reference transcriptome assembly.

Transcriptome Analysis

Gene expression profiling by RNA-Seq provides an unprecedented high-resolution view of the global transcriptional landscape. As the sequencing technologies and protocol methodologies continually evolve, new informatics challenges and applications develop. Beyond surveying gene expression levels, RNA-Seq can also be applied to discover novel gene structures, alternatively spliced isoforms, and allele-specific expression (ASE). In addition, genetic studies of gene expression using RNA-Seq have observed genetically correlated variability in expression, splicing, and ASE.

RNA-Sequencing Data Analysis Workflow

The conventional pipeline for RNA-Seq data includes generating FASTQ-format files contains reads sequenced from an NGS platform, aligning these reads to an annotated reference genome, and quantifying expression of genes. Although basic sequencing analysis tools are more accessible than ever, RNA-Seq analysis presents unique computational challenges not encountered in other sequencing-based analyses and requires specific consideration to the biases inherent in expression data.

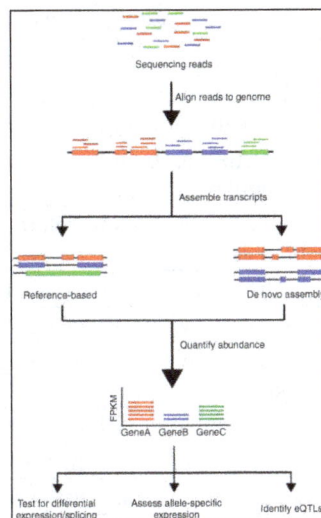

Overview of RNA-Seq data analysis. Following typical RNA-Seq experiments, reads are first aligned to a reference genome. Second, the reads may be assembled into transcripts using reference transcript annotations or de novo assembly approaches. Next, the expression level of each gene is estimated by counting the number of reads that align to each exon or full-length transcript. Downstream analyses with RNA-Seq data include testing for differential expression between samples, detecting allele-specific expression, and identifying expression quantitative trait loci (eQTLs).

Read Alignment

Mapping RNA-Seq reads to the genome is considerably more challenging than mapping DNA sequencing reads because many reads map across splice junctions. In fact, conventional read mapping algorithms, such as Bowtie and BWA, are not recommended for mapping RNA-Seq reads to the reference genome because of their inability to handle spliced transcripts. One approach to resolving this problem is to supplement the reference genome with sequences derived from exon–exon splice junctions acquired from known gene annotations. A preferred strategy is to map reads with a "splicing-aware" aligner that can recognize the difference between a read aligning across an exon–intron boundary and a read with a short insertion. As RNA-Seq data have become more widely used, a number of splicing-aware mapping tools have been developed specifically for mapping transcriptome data. The more commonly used RNA-Seq alignment tools include GSNAP, MapSplice, RUM,and TopHat. Each aligner has different advantages in terms of performance, speed, and memory utilization. Selecting the best aligner to use depends on these metrics and the overall objectives of the RNA-Seq study. Efforts to systematically evaluate the performance of RNA-Seq aligners have been initiated by GENCODE's RNA-Seq Genome Annotation Assessment Project3 (RGASP3), which has found major performance difference between alignments tools on numerous benchmarks, including alignment yield, basewise accuracy, mismatch and gap placement, and exon junction discovery.

Table: Widely used RNA-Seq software packages

Primary category	Tool name	Notes
Splice-aware read alignment	GEM	Filtration-based approach to approximate string matching for alignment
	GSNAP	Based on seed and extend alignment algorithm aware of complex variants
	MapSplice	Based on Burrows-Wheeler Transform (BWT) algorithm
	RUM	Integrates alignment tools Blat and Bowtie to increase accuracy
	STAR	Based on seed searching in an uncompressed suffix arrays followed by seed clustering and stitching procedure; fast but memory-intensive
	TopHat	Uses Bowtie, based on BWT, to align reads; resolves spliced reads using exons by split read mapping
Transcript assembly and quantification	Cufflinks	Assembles transcripts to reference annotations or de novo and quantifies abundance
	FluxCapacitor	Quantifies transcripts using reference annotations
	iReckon	Models novel isoforms and estimates their abundance
Differential expression (DE)	BaySeq	Count-based approach using empirical Bayesian method to estimate posterior likelihoods
	Cuffdiff2	Isoform-based approach based on beta negative binomial distribution
	DESeq	Exon-based approach using the negative binomial model
	DEGSeq	Isoform-based approach using the Poisson model
	EdgeR	Count-based approach using empirical Bayes method based on the negative binomial model
	MISO	Isoform-based model using Bayes factors to estimate posterior probabilities
Other tools	HCP	Normalizes expression data by inferring known and hidden factors with prior knowledge
	PEER	Normalizes expression data by inferring known and hidden factors using a probabilistic estimation based on the Bayesian framework
	Matrix eQTL	Fast eQTL detection tool that uses linear models (linear regression or ANOVA)

Transcript Assembly and Quantification

After RNA-Seq reads are aligned, the mapped reads can be assembled into transcripts. The majority of computational programs infer transcript models from the accumulation of read alignments to the reference genome. An alternative approach for transcript assembly is de novo reconstruction, in which contiguous transcript sequences are assembled with the use of a reference genome or annotations. The reconstruction of transcripts from short-read data is a major challenge and a gold standard method for transcript assembly does not exist. The nature of the transcriptome (e.g., gene complexity, degree of polymorphisms, alternative splicing, dynamic range of expression), common technological challenges (e.g., sequencing errors), and features of the bioinformatics workflow (e.g., gene annotation, inference of isoforms) can substantially affect transcriptome assembly quality. RGASP3 has initiated efforts to evaluate computational methods for transcriptome reconstruction and has found that most algorithms can identify discrete transcript components, but the assembly of complete transcript structures remains a major challenge.

A common downstream feature of transcript reconstruction software is the estimation of gene expression levels. Computational tools such as Cufflinks, FluxCapacitor, and MISO, quantify expression by counting the number of reads that map to full-length transcripts. Alternative approaches, such as HTSeq, can quantify expression without assembling transcripts by counting the number of reads that map to an exon. To accurately estimate gene expression, read counts must be normalized to correct for systematic variability, such as library fragment size, sequence composition bias, and read depth. To account for these sources of variability, the reads per kilobase of transcripts per million mapped reads (RPKM) metric normalizes a transcript's read count by both the gene length and the total number of mapped reads in the sample. For paired end-reads, a metric that normalizes for sources of variances in transcript quantification is the paired fragments per kilobase of transcript per million mapped reads (FPKM) metric, which accounts for the dependency between paired-end reads in the RPKM estimate. Another technical challenge for transcript quantification is the mapping of reads to multiple transcripts that are a result of genes with multiple isoforms or close paralogs. One solution to correct for this "read assignment uncertainty" is to exclude all reads that do not map uniquely, as in Alexa-Seq. However, this strategy is far from ideal for genes lacking unique exons. An alternative strategy used by Cufflinks, and MISO is to construct a likelihood function that models the sequencing experiment and estimates the maximum likelihood that a read maps to a particular isoform.

Considerations for miRNA Sequencing Analysis

The general approach for analysis of miRNA sequencing data is similar to approaches discussed for mRNA. To identify known miRNAs, the sequencing reads can be mapped to a specific database, such as miRBase, a repository containing over 24,500 miRNA loci from 206 species in its latest release (v21) in June 2014. In addition, several tools have been developed to facilitate analysis of miRNAs including the commonly used tools miRanalyzer and miRDeep. MiRanalyzer can detect known miRNAs annotated on miRBase as well as predict novel miRNAs using a machine-learning approach based on the random forest method with a broad range of features. Similarly, miRDeep is able to identify known miRNAs and predict novel miRNAs using properties of miRNA biogenesis to score the compatibility of the position and frequency of sequenced RNA from the secondary structure of precursor miRNAs. Although miRDeep and miRanalyzer contain modules for target prediction, expression quantification, and differential expression, the methods developed for mRNA quantification and differential expression can also be applied to miRNA data.

Quality Assessment and Technical Considerations

At each stage in the RNA-Seq analysis pipeline, careful consideration should be applied to identifying and correcting for various sources of bias. Bias can arise throughout the RNA-Seq experimental pipeline, including during RNA extraction, sample preparation, library construction, sequencing, and read mapping. First, the quality of the raw sequence data in FASTQ-format files should be evaluated to ensure high-quality reads. User-friendly software tools designed to generate quality overviews include the FASTX-toolkit, the FastQC software, and the RobiNA package. Several important parameters that should be evaluated include the sequence diversity of reads, adaptor contamination, base qualities, nucleotide composition, and percentage of called bases. These technical artifacts can arise at the sequencing stage or during the construction of the RNA-Seq. For example, the 5′ read end, derived from either end of a double-stranded cDNA fragment, shows higher error rate due to mispriming events introduced by the random oligos during the RNA-Seq library construction protocol. If possible, actions to correct for these biases should be performed, such as trimming the ends of reads, to expedite the speed and improve the quality of the read alignments.

After aligning the reads, additional parameters should be assessed to account for biases that arise at the read mapping stage. These parameters include the percentage of reads mapped to the transcriptome, the percentage of reads with a mapped mate pair, the coverage bias at the 5′- and 3′-ends, and the chromosomal distribution of reads. One of the most common sources of mapping errors for RNA-Seq data occurs when a read spans the splicing junction of an alternatively spliced gene. A misalignment can be easily introduced due to ambiguous mapping of the read end to one of the two (or more) possible exons and is especially common when reads are mapped to a reference transcriptome that contains an incomplete annotation of isoforms. If genotype information is available, the integrity of the samples should also be evaluated by investigating the correlation of single-nucleotide variants (SNVs) between the DNA and RNA reads. The concordance between the DNA and RNA sequencing data may provide insight into sample swaps or sample mixtures caused accidentally as a result of personnel or equipment error. In the case of a swapped sample, more discordant variants would be observed between the DNA and RNA sequencing data. In the case of a mixture of samples, more significant patterns of allele-specific expression would be observed than expected for a single individual as a result of more combinations of heterozygous and homozygous sites that would skew the alleles beyond the expected 1:1 allelic ratio.

Differential Gene Expression

A primary objective of many gene expression experiments is to detect transcripts showing differential expression across various conditions. Extensive statistical approaches have been developed to test for differential expression with microarray data, where the continuous probe intensities across replicates can be approximated by a normal distribution. Although in principle these approaches are also applicable to RNA-Seq data, different statistical models must be considered for discrete read counts that do not fit a normal distribution. Early RNA-Seq studies suggested that the distribution of read counts across replicates fit a Poisson distribution, which formed the basis for modeling RNA-Seq count data. However, further studies indicated that biological variability is not captured by the Poisson assumption, resulting in high false-positive rates due to underestimation of sampling error. Hence, negative binomial distribution models that take into account overdispersion or extra-Poisson variation have been shown to best fit the distribution of read counts across biological replicates.

To model the count-based nature of RNA-Seq data, complex statistical models have been developed to handle sources of variability that model overdispersion across technical and biological replicates. One source of variability is differences in sequencing read depth, which can artificially create differences between samples. For instance, differences in read depth will result in the samples appearing more divergent if raw read counts between genes are compared. To correct for this, it is advantageous to transform raw read count data to FPKM or RPKM values in differential expression analyses. Although this correction metric is commonly used in place of read counts, the presence of several highly expressed genes in a particular sample can significantly alter the RPKM and FPKM values. For example, a highly expressed gene can "absorb" many reads, consequently repressing the read counts for other genes and artificially inflating gene expression variation. To account for this bias, several statistical models have been proposed that use the highly expressed genes as model covariates. Another source of variability that has been observed is that the distribution of sequencing reads is unequal across genes. Therefore, a two-parameter generalized Poisson model that simultaneously considers read depth and sequencing bias as independent parameters was developed and shown to improve RNA-Seq analysis. More complex normalization methods have also been developed to account for hidden covariates without removing significant biological variability. For example, the probabilistic estimation of expression residuals (PEER) framework and the hidden covariates with prior (HCP) framework are methods that use a Bayesian approach to infer hidden covariates and remove their effects from expression data.

To detect differential expression, a variety of statistical methods have been designed specifically for RNA-Seq data. A popular tool to detect differential expression is Cuffdiff, which is part of the Tuxedo suite of tools (Bowtie, Tophat, and Cufflinks) developed to analyze RNA-Seq data. In addition to Cuffdiff, several other packages support testing differential expression, including bay-Seq, DESeq, DEGseq, and edgeR. Although these packages can assign significance to differentially expressed transcripts, the biological observations should be carefully interpreted. Each model makes specific assumptions that may be violated in the context of the observed data; therefore, an understanding of the model parameters and their constraints is critical for drawing meaningful and accurate biological conclusions. Furthermore, replicates in RNA-Seq experiments are crucial for measuring variability and improving estimations for the model parameters. Biological replicates (e.g., cells grown on two different plates under the same conditions) are preferred to technical replicates (e.g., one RNA-Seq library sequenced on two different lanes), which show little variation. Although the number of replicates required per condition is an open research question, a minimum of three replicates per sample has been suggested. In many cases, multiplexed RNA-Seq libraries can be used to add biological replicates without increasing sequencing costs (if sequenced at a lower depth) and will greatly improve the robustness of the experimental design. Additionally, the accuracy of measurements of differential gene expression can be further improved by using ERCC spike-in controls to distinguish technical variation from biological variation.

Allele-Specific Expression

A major advantage of RNA-Seq is the ability to profile transcriptome dynamics at a single-nucleotide resolution. Therefore, the sequenced transcript reads can provide coverage across heterozygous sites, representing transcription from both the maternal and paternal alleles. If a sufficient number of reads cover a heterozygous site within a gene, the null hypothesis is that the ratio of maternal to paternal alleles is balanced. Significant deviation from this expectation suggests allele-specific expression (ASE). Potential mechanisms for ASE include genetic variation (e.g., single-nucleotide polymorphism in a *cis*-regulatory region upstream of a gene) and epigenetic effects (e.g., genomic imprinting, methylation, histone modifications, etc.). Early studies showed that allele-specific differences can affect up to 30% of loci within an individual and are caused by both common and rare genetic variants. Studies have also applied ASE to identify expression modifiers of protein-coding variation, effects of loss-of-function variation, and differences between pathogenic and healthy tissues. Furthermore, ASE studies using single-cell transcriptomics have uncovered a stochastic pattern of allelic expression that may contribute to variable expressivity, a novel perspective which may have fundamental implications for variable disease penetrance and severity.

Conventional workflows to detect ASE involve counting reads containing each allele at heterozygous sites and applying a statistical test, such as the binomial test or the Fisher's exact test. However, more rigorous statistical approaches are necessary to overcome technical challenges involved in ASE detection. These challenges include read-mapping bias, sampling variance, overdispersion at extreme read depths, alternatively spliced alleles, insertions and deletions (indels), and genotyping errors. To account for overdispersion, one approach is to model allelic read counts using a beta-binomial distribution at individual loci; however, accurate estimation of the overdispersion parameter requires replicates and, in our experience, major source of bias come from site-specific mapping differences. Another strategy is to use a hierarchical Bayesian model that combines information across loci, as well as across replicates and technologies, to make global and site-specific inferences for ASE. To assess reference-allele mapping bias, the number of mismatches in reads containing the nonreference allele should be assessed as increased bias is observed with greater sequence divergence between alleles. To correct for read-mapping bias, an enhanced reference genome can be constructed that masks all SNP positions or includes the alternative alleles at polymorphic loci. Statistical methods to better address these technical biases are under active development and are expected to foster further improvements in ASE detection.

Expression Quantitative Trait Loci

Another prominent direction of RNA-Seq studies has been the integration of expression data with other types of biological information, such as genotyping data. The combination of RNA-Seq with genetic variation data has enabled the identification of genetic loci correlated with gene expression variation, also known as expression quantitative trait loci (eQTLs). This expression variation caused by common and rare variants is postulated to contribute to phenotypic variation and susceptibility to complex disease across individuals. The goal of eQTL analysis is to identify associations that will uncover underlying biological processes, discover genetic variants causing disease, and determine causal pathways. Initial eQTL studies using RNA-Seq data identified a greater number of statistically significant eQTLs than had been identified by microarray studies. Most of the eQTLs identified directly influenced gene expression in an allele-specific manner and were located near transcriptional start sites, indicating that eQTLs could modulate expression directly, or in cis. Later studies identified *trans*-eQTLs, which are variants that affect the expression of a distant gene (>1 Mb) by modifying the activity or expression of upstream factors that regulate the gene. Although *trans*-eQTLs show weaker effects and present validation difficulties, they can potentially reveal previously unknown pathways in gene regulation networks.

RNA-Seq has revolutionized QTL analyses because it enables association analyses of more than just gene expression levels alone. For example, RNA-Seq provides unprecedented opportunity to investigate variations in splicing by profiling alternately spliced isoforms of a gene. This has enabled the identification of variants influencing the quantitative expression of alternatively spliced isoforms commonly referred to as splicing-QTLs (sQTLs). In addition, specific RNA-Seq library constructions (e.g., ribo-depleted) have enabled the detection of eQTLs affecting other RNA species; recent studies have identified variants affecting the expression of various ncRNAs, including long intergenic noncoding RNAs. The expanding potential of RNA-Seq to associate phenotypic variations with genetic variation offers an enhanced understanding of gene regulation.

Traditional eQTL mapping methods that were developed for microarray data use linear models such as linear regression and ANOVA to associate genetic variants with gene expression. These methods have been directly applied to RNA-Seq data following appropriate normalization of total read counts. Most eQTL studies perform separate testing for each transcript-SNP pair using linear regression and ANOVA models to detect significant association. Nonlinear approaches have also been developed to test associations, such as generalized linear and mixed models, Bayesian regression. Alternative models, such as Merlin, have also been developed to detect eQTLs from expression data that include related individuals using pedigree data. In addition, several methods have been developed to simultaneously test the effect of multiple SNPs on the expression of a single gene using Bayesian methods. To further improve on the detection of causal regulatory variants, several studies have integrated ASE information with eQTL analysis. These studies showed that genetic variants showing allele-specific effects and identified as eQTLs show higher enrichment in functional annotations and provide stronger evidence of *cis*-regulatory impact. Because high-throughput sequencing has created genotype data sets featuring millions of SNPs and expression data sets featuring tens of thousands of transcripts, the task of testing billions of transcript-SNP pairs in eQTL analysis can be computationally intensive. To mitigate this computational burden, software has been developed such as Matrix eQTL to efficiently test the associations by modeling the effect of genotype as either additive linear (least squares model) or categorical (ANOVA model). Because of the large number of tests performed, it is important to correct for multiple-testing by calculating the false discovery rate or resampling using bootstrap or permutation procedures.

However, the design and interpretation of eQTL studies is not straightforward. Many complications result from the complexity of gene regulation, which shows both spatial (cell and tissue location) specificity as well as temporal (developmental stage) specificity. For instance, several studies have performed eQTL analysis across multiple tissues, indicating that genetic regulatory elements can have tissue-specific effects. Therefore, future eQTL analyses should test for SNP-transcript associations in well-defined cell types that are relevant to the trait of interest. For example, a study detecting eQTLs in cardiovascular disease should use heart tissue while a study interested in autoimmune disease should use whole blood. Another major consideration for eQTL studies is accounting for population structure and elucidating the causal variants. The structure of genomic variation can vary significantly between populations and will influence the resolution of any genetic association study. Furthermore, if substantial linkage disequilibrium (LD) exists within the genome, the associated genetic variant is often "tagging" the causal variant rather than acting as the causal regulatory variant itself. As eQTL studies integrate data across different populations and use population-scale genome sequencing, the ability to elucidate causal variants will greatly improve.

References

- DNA-sequencing, science: britannica.com, Retrieved 5 May, 2019

- Dna-sequence-data-analysis-starting-off-in-bioinformatics: towardsdatascience.com, Retrieved 10 March, 2019

- Introduction-gene-expression-profiling, gene-expression-analysis, life-science: thermofisher.com, Retrieved 21 January, 2019

- What-gene-expression-analysis: bio-rad.com, Retrieved 17 July, 2019

- Serial-Analysis-of-Gene-Expression, life-sciences: news-medical.net, Retrieved 3 May, 2019

- Whole-genome-sequencing, dna-sequencing, sequencing: illumina.com, Retrieved 27 February, 2019

- Whole-genome-sequencing-methods, hematology-oncology: healio.com, Retrieved 29 August, 2019

- What-is-gene-annotation-in-bioinformatics: biolyse.ca, Retrieved 19 April, 2019

- Transcriptome-sequencing: allseq.com, Retrieved 14 January, 2019

Chapter 5

Proteome and Protein Sequence Analysis

The methods of determining the amino acid sequence of protein or peptide is known as protein sequencing. The entire set of proteins which can be expressed by an organism is known as proteome. Its identification as well as quantification is called proteomic analysis. This chapter discusses in detail the theories and methodologies related to proteome and protein sequence analysis.

Protein Sequencing

The molecules that give cells and entire organisms their shape as well as their ability to move, grow, and reproduce are the proteins. Although they come in an almost infinite variety of shapes and sizes, they have all been designed by the process of evolution to serve a defined and useful function in the processes of life. Some proteins, like actin and collagen, help to give a cell its physical shape. Other proteins, like lactase and pepsin, help in the digestion of food. Others transport signals between cells, help us fight off disease, or repair damaged DNA. For almost every job in a cell, there is a protein designed to do it.

The Building Blocks of Proteins

The building blocks of proteins are amino acids. There are twenty different amino acids used by living cells to build proteins. They are linked together in a long, linear chain during the process of translation, which is carried out by the ribosomes inside cells. Proteins begin to take on their characteristic three-dimensional shape even while they are being made, folding and twisting as each new amino acid added to the chain tugs or pushes at the others added before it. Each amino acid has an amino group ($-NH_3^+$) and a carboxyl group (-COOH). Peptide bonds link the carboxyl group of one amino acid to the amino group of the next amino acid. On one end of a protein, therefore, there is a free amino group called the N-terminus, and on the other end is a free carboxyl group, called the C-terminus.

The process of determining a protein's order of amino acids is called protein sequencing. A protein's sequence can easily be deduced from its gene sequence, since the order of bases on a DNA strand specifies the order in which the amino acids are linked together during translation. The chemistry involved in DNA sequencing is less complex than that which is involved in determining the order of each amino acid in an amino acid chain. There are two primary reasons why effort would be put into sequencing a protein. The first is to provide the information needed to design a synthetic DNA probe that can be used to locate the gene that codes for the protein. The second is to prove that a protein that has been isolated or manufactured in the laboratory is what it is believed to be.

Sequencing Techniques

The most widely used technique for sequencing proteins is the Edman degradation, a procedure developed by Pehr Edman in the 1950s. The reaction steps used for this method have since been completely automated by machine. The procedure uses special reagents under alternating basic and acidic conditions to remove one amino acid at a time from the protein's N-terminus. As each amino acid is released during each cycle of degradation, it is identified by chromatography, a separation technique that relies on an amino acid's unique size and electrical charge to distinguish it from the other nineteen amino acids.

In many automated approaches, high-performance liquid chromatography (HPLC) is used to tell which amino acid has been released; the amount of time it takes to travel through an HPLC column is unique to each amino acid. Up to fifty amino acids from the N-terminus can be identified using Edman degradation. If a scientist is trying to identify a previously sequenced protein, usually only the first fifteen to twenty amino acids of the purified protein need to be sequenced. That information can then be entered into a database and matched with known proteins having identical or related sequences.

Sequencing a protein from its C-terminus is particularly challenging, and there are no techniques that are as robust as Edman degradation. However, some limited amino acid sequence information can be obtained using enzymes called carboxypeptidases, which remove individual C-terminal amino acids. These enzymes, however, tend to cleave only specific amino acids from the C-terminus.

Carboxypeptidase B, isolated from cow pancreas, for example, can release the amino acids arginine and lysine from the C-terminus of a protein. Carboxypeptidase A, also isolated from cow pancreas, fails to release arginine, lysine, or proline, but can cleave off the other seventeen amino acids. Carboxypeptidases isolated from citrus leaves and yeast can cleave off any amino acid from the C-terminus of a protein, although the rate at which they do this depends on the particular amino acid. If one amino acid is released slowly and the next within the chain is released very quickly, they might appear to be cleaved at the same time, making it difficult to establish their order. C-terminal amino acid identification using enzymes, therefore, is not practical beyond the first several positions.

Another method of protein sequencing, called mass spectrometry, uses electric current to break individual amino acids from a protein. In a mass spectrometer, the released amino acids are collected in a detector and are each identified by their unique mass.

Sequencing of the human genome has allowed a giant leap in the understanding of how the human species evolved and how genetic diseases arise. Advances made in DNA sequencing technology lead to this grand accomplishment. The next frontier is to decipher how all the proteins encoded by the genome interact to carry out the processes of life. This is the study of proteomics. Advances in mass spectrometry and protein sequencing instrumentation are bringing this challenging problem closer to its resolution.

Protein Sequence Alignment and Analysis

Amino acid sequence alignment and analysis is central to most biochemical and molecular biology applications. Although it should be possible to retrieve all the information we need about a protein directly from its sequence, looking at a sequence without prior knowledge and experience is like reading a text in a foreign language: we may recognize the letters, but we do not understand the meaning and are unable to extract the information. Still, when proteins are concerned, we have learned to extract a substantial part of the information from detailed sequence analysis, using for example multiple sequence alignment. In a multiple sequence alignment a given sequence is compared to a group of other sequences from related organisms. When we say "related" we mean "evolutionary related" and that they belong to the same family, the members of which usually perform a similar function in different organisms. We know that when proteins are evolutionary related the main characteristic features of the sequence and the tertiary structure are conserved. Since conservation of function normally assumes that a certain number of amino acid residues within a protein family are conserved, we need to have some tools to be able to assess the degree of conservation of each member of the protein family. For this, alignment techniques and scoring schemes for sequence alignment have been developed.

In a sequence alignment we try to align identical amino acids in the sequences against each other. However, since normally there are also many amino acid substitutions, we need to know how to handle substitutions of one amino acid by another in the sequences being aligned (amino acid substitutions are caused by mutations in the gene coding for the protein in question). Some substitutions are conservative, i.e., they will not cause any substantial disturbances in the protein structure, which would affect the protein function, but other substitutions, if they would occur, may have a dramatic effect on protein structure and function. To handle amino acid substitutions in sequence alignment, specially designed substitution matrices are used, which are part of the alignment scoring scheme and help in calculating the score of the alignment to distinguish between several possible alignments. Even structural information may be used in making a correct alignment, for example in correctly placing insertion and deletion regions in the alignment. Insertions and deletions are very common in sequences belonging to the same family and often occur in loop regions. By other words, insertions and deletions may indicate that a certain region of the sequence may have a loop structure.

Amino Acid Analysis

Amino acid analysis is a fundamental biochemical technique used for the determination of the amino acid composition or content of proteins, peptides and other pharmaceutical or biological preparations or samples containing compounds that contain primary or secondary amino groups within their molecular structure. Amino acid analysis allows for amino acid quantitation (also known as amino acid quantification or amino acid identification) of free amino acids, as well as amino acids released from macromolecules such as peptides, proteins or glycoproteins. Amino acid testing also enables the analysis of protein complexes or mixtures of proteins such as protein powder supplements.

Protein molecules are abundant in mammals and are a significant and vital part of the mammalian diet as well as a vital part of their metabolism. Since proteins and various amino acids are needed in the human diet to help the body repair cells and synthesize new cells, amino acid quantification may be used to monitor or detect the metabolic states by analyzing the content of free amino acids in biological fluids such as urine, blood or plasma.

Methods for Amino Acids Analysis

High Performance Liquid Chromatography is the most popular approach for amino acid components analysis. However, there are many types of amino acids analysis.

Detection Types

Using UV detection for amino acids in most cases requires by using the absorption of the carboxyl group in the 200 to 210 nm range. In generally, some amino acids are difficult to analyze as-is with sufficient sensitivity and selectivity. Finally, derivatization methods, which include many types, have been used for a long time.

Pre-label Method

As for this method, the amino acids are derivertized before injection and then the reaction products are separated and detected. There are some advantages of using this method:

1. Reagent consumption rates can be minimized by specifying a small reaction system.

2. It allows increasing sensitivity by using more expensive reagents, which offer lower background levels.

3. Even if unreacted derivatizing reagent is detected, it will not be a problem as long as it is separated in the column.

Disadvantages of pre-label method:

1. Sample matrix will affect derivatization reaction efficiency.

2. Reaction products are often unstable and may affect quantitation results.

Post-column Reaction Detection Method

There are some steps in post-column derivatization method, including amino acids separation, derivatizing reagent delivery and mix, which can allow it to react with the amino acids before sending the products to the detector. Some advantages of this method are listed below:

1. It can be automated offering excellent quantitative performance and reproducibility.

2. Due to the separation of sample components before reaction, reaction efficiency if less prone to sample matrix effects, which enable it to be applied for a wide range of samples.

Disadvantages of post-column derivatization:

1. It is not suitable for high-sensitivity analysis.

2. Relatively high reagent consumption.

3. Limited variety of derivatizing reagents will be applied.

4. Rapid reversed-phase chromatography can not be used.

Protein Structure Prediction Methods

Constituent amino-acids can be analyzed to predict secondary, tertiary and quaternary protein structure.

Protein structure prediction is the prediction of the three-dimensional structure of a protein from its amino acid sequence — that is, the prediction of its folding and its secondary, tertiary, and quaternary structure from its primary structure. Structure prediction is fundamentally different from the inverse problem of protein design. Protein structure prediction is one of the most important goals pursued by bioinformatics and theoretical chemistry; it is highly important in medicine (for example, in drug design) and biotechnology (for example, in the design of novel enzymes).

There are three major theoretical methods for predicting the structure of proteins: comparative modelling, fold recognition, and ab initio prediction.

Comparative Modelling

Comparative modelling exploits the fact that evolutionarily related proteins with similar sequences, as measured by the percentage of identical residues at each position based on an optimal structural superposition, have similar structures. The similarity of structures is very high in the so-called core regions", which typically are comprised of a framework of secondary structure elements such as alpha-helices and beta-sheets. Loop regions connect these secondary structures and generally vary even in pairs of homologous structures with a high degree of sequence similarity.

Fold Recognition or Threading

Threading uses a database of known three-dimensional structures to match sequences without known structure with protein folds. This is accomplished by the aid of a scoring function that assesses the fit of a sequence to a given fold. These functions are usually derived from a database of known structures and generally include a pairwise atom contact and solvation terms. Threading methods compare a target sequence against a library of structural templates, producing a list of scores. The scores are then ranked and the fold with the best score is assumed to be the one adopted by the sequence. The methods to fit a sequence against a library of folds can be extremely elaborate computationally, such as those involving double dynamic programming, dynamic programming with frozen approximation, Gibbs Sampling using a database of "threading" cores, and branch and bound heuristics, or as "simple" as using sophisticated sequence alignment methods such as Hidden Markov Models.

Ab Initio Prediction

The ab initio approach is a mixture of science and engineering. The science is in understanding how the three-dimensional structure of proteins is attained. The engineering portion is in deducing the three-dimensional structure given the sequence. The biggest challenge with regards to the folding problem is with regards to ab initio prediction, which can be broken down into two components: devising a scoring function that can distinguish between correct (native or native-like) structures from incorrect (non-native) ones, and a search method to explore the conformational space. In many ab initio methods, the two components are coupled together such that a search function drives, and is driven by, the scoring function to find native-like structures.

Protein Complex Analysis

Proteins, also called polypeptides, are the polymers of amino acids. There are a total of twenty amino acids called monomers that exist naturally in proteins. Proteins are found in abundance and are differentiated from each other according to the number, type, and arrangement of amino acids in series, which comprise the mainstay of polypeptides.

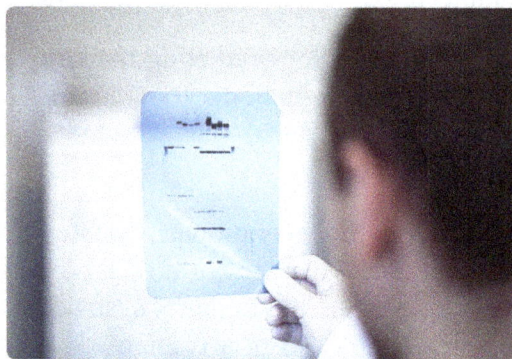

For different reasons, proteins are the main components of food, as well as an important source for producing energy. Proteins are generally present in many natural food types such as fish or meat. Many of the proteins in food are enzymes that are able to improve a certain number of biochemical reactions. The presence of amino acids in proteins is very essential for human health.

Methods Involved in Analyzing Protein

Different proteins have different molecular structures, physiochemical properties, and nutritional attributes. Three major steps are involved in protein analysis.

1. Protein separation

The following are some of the common methods for separating proteins:

- SDS (sodium dodecyl sulfate)-PAGE: Protein separation by this method is based on the molecular weight, as opposed to fold or charge. Factors other than the molecular weight, however, may also affect the movement of proteins on SDS gel, which was analyzed in a course group. This technique is broadly used in genetics, biochemistry, molecular biology, and forensics.

- Isoelectric focusing: Separation of protein by this method is dependent upon balance between the negative charged (acidic) and positive charged (basic) constituent amino acids, which defines a property called protein isoelectric (PI) point. The net charge of the proteins in pH becomes accurately zero. Proteins that vary by as small as a single charged residue can be isolated by this method.

On the other hand, however, proteins of different sizes with a similar PI value can be placed at the same place. Often, two chromatic methods are used for the separation of proteins or even for peptide separation.

- High-performance liquid chromatography (HPLC) methods are used for separating and purifying peptides or proteins on the basis of charge, size, or on the whole hydrophobicity.

- Thin-layer chromatography (TLC) has also been used for separating peptides on the basis of their similar properties.

- Gel Electrophoresis through two-dimensions (2D): This is a powerful technique generally used for analyzing complex samples, but it may not be used for a few or specific proteins, in the interest of characterizing all varieties of protein samples. In this method, first the protein is passed in a narrow path of isoelectric targeting gel. Based on the isoelectric point the protein can be separated.

Then the isoelectric targeting gel is situated over the SDS-PAGE gel and it is run in the perpendicular direction. Proteins run as spots and their position is based on their size and charged properties. This provides a powerful separation of proteins using either of these two gels only, which are isoelectric focusing or SDS-PAGE.

2. Western blotting

Immunoblotting is considered the most general version of western blotting and is used in combination with chromatographic methods. This method is used for identifying specific proteins within a given piece of tissue extract or homogenate. The protein sample is first electrophoresed by using SDS page for the separation of proteins on the basis of molecular weight.

Following this, the wet gel is positioned against the nitrocellulose sheet and kept in a special kind of electrophoretic chamber. Thereafter, an electric field is passed into the gel that causes the protein to move out of the gel and also onto the sheet of nitrocellulose in which it becomes adsorbed tightly. Afterward, nitrocellulose sheet with this tightly bounded protein can be "probed" or "blotted" with antibodies, specifically for targeting the protein.

3. Protein identification

If the protein spot or band was found on a gel, the next step is to determine their molecular identity. The two methods commonly available for this method are as follows:

- Edman degradation: This method was developed by Pehr Edman. He used reagents called isothiocyanates to act and react with the N-terminus of the peptide or protein. Under certain suitable conditions, a solitary "round" of Edman degradation can split the amino terminus residue to produce amino acid derivative with the addition of a new free N-terminus. Eventually, the amino terminus is separated from the protein, without interrupting the bonding of proteins between the other residual amino acids.

- Mass spectrometry: This is the most apt method for identifying the exact protein molecular weight. Because this method is very sensitive and amenable to the quick processing of several samples, this technique has become popular for identifying the different kinds of proteins that are present in a complex sample. Hence, mass spectrometry may be used for determination of the sequence of the proteins and thus the protein identity in a way conceptually like the above described, based on an Edman degradation.

Challenges of Protein Complex Analysis

Protein complex analysis involves extensive interpretation of the structure and function of proteins, which are present in complex biological samples. Due to this, techniques like protein complex analysis have great value in understanding the complex organisms. Though recent protein complex analysis methods are efficient in identifying the structure and of protein complex, there are some limiting factors.

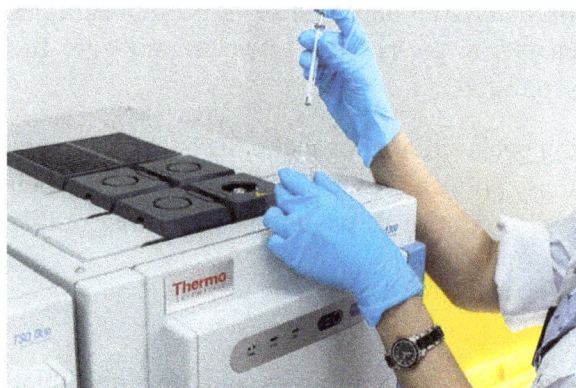

Some of the factors that affect protein complex analysis are serum proteomics, the presence of membrane protein in proteome, complexity, etc. Much of the research is being performed to enhance the protein complex analysis process, which includes identifying whether the complexes are suitable for being modulated efficiently by small molecules.

Interference of Protein–Protein Interactions

One of the main challenges faced during the protein complex analysis process is the development of compounds that interfere by acting on protein complexes.

In some cases like those involving ion channels (considered as the reliable target class earlier), the process of directing the compounds against individual proteins has not been possible whereas hist is implemented in many other cases. Simultaneously, several compounds have been identified, which act on alpha subunits involved in pore formation.

Some of those compounds are successfully implemented in clinical studies, which is mainly due to their critical side effects and lack of specificity. Many compounds that had been developed to act against individual proteins are now being implemented in acting against other targets.

Even at present, there exists a misconception that the interference of small molecules with protein–protein interactions is not possible, which is mainly due to the lack of understanding regarding the protein–protein interactions process. Protein interfaces have conformational flexibility in constitutive complex too. Moreover, studies involving allosteric proteins indicate that the ligands are responsible for structural changes that occur at protein–protein interfaces.

The binding sites of allosteric ligands are not restricted sterically, which is the advantage, and indicates that they are as efficient as traditional competitive agonist. Allosteric ligands are efficient in developing drugs that can act against protein complexes. The cell-based assays and high-content screening are the techniques involved in developing drugs against protein complexes.

Complexity Challenge

The complexity challenge is another factor that affects the analysis of protein complexes. Two important strategies that are involved in resolving these challenges are recombinant approaches and those that are based on native sources.

Recombinant approaches, such as tagged protein co-purification analysis and yeast-two hybrid screens, consist of a systematic methodology that makes them reliable to resolve the complexity challenge. Both the aforementioned are based on gene expression from random libraries of yeast. Those methods prove to be inefficient when applied to proteome of simple organism. Half of the interactions are false-positive and interactions that involve membrane proteins are not detected.

Approaches that are based on native sources are based on both biochemical fractionation and mass-spectrometric identification. Some of the restrictions in these methods are complex source preparation methods and the amount of sample required.

The biochemical fractionation approach is an option for affinity purification based on a protein. Requirement of high affinity ligands is also a limiting factor in these approaches. High sensitivity and mass spectrometry reliability should be efficient as those in recent bioinformatics techniques. Aforementioned limitations restrict the large-scale application of these methods. Nevertheless, better results can be obtained if these approaches are carried out perfectly.

Presence of Membrane Proteins

Membrane proteins occupy over 30% of the entire complement of protein. Unfortunately, their tendency to get aggregated and get deposited in solid form in a solution restricts their analysis. Tryptic cleavage target residues such as arginine and lysine are not present in transmembrane segments, but are present only in the hydrophilic part of these membrane proteins.

Two-dimensional gel electrophoresis is not suitable for isolating these membrane proteins. These challenges can be overcome by various membrane solubilizing approaches that are used in resolving the solubility issues and in the analysis of fractions enriched in membranes.

In this method, solubilization is achieved through various processes, such as a fraction of an enriched yeast membrane is subjected to solubilization process in the existence of cyanogen bromide by adding 90% of formic acid solution; a microsomal membrane fraction is subjected to solubilization by boiling it in 0.5% of sodium dodecyl sulfate solution; and by using a sample of enriched membrane in thermal sonication and denaturation processes of proteins in 60% of methanol solution with trypsin. All the solubilization methods discussed earlier are efficient and used in optimization of membrane protein identification process.

Serum Proteomics

Human serum contains about 10,000 types of proteins. The large variation in the serum concentration was a limiting factor for biomarker discovery. Human serum concentration contains a high amount of albumin when compared to other protein types but eliminating the albumin may also result in the elimination of other types of proteins such as peptide hormones, cytokines, etc. Various methods are currently used, which are efficient in identifying biomarkers discovered from serum. Some of the methods include 2D-PAGE analysis, SELDI-TOF MS, etc.

References

- Amino-acid-sequencing, genetics-and-genetic, biology-and-genetics: encyclopedia.com, Retrieved 5 June, 2019
- Sequence-analysis, Sequence: proteinstructures.com, Retrieved 15 January, 2019
- Amino-acid-analysis: biosyn.com, Retrieved 17 July, 2019
- Methods-for-amino-acids-analysis: creative-proteomics.com, Retrieved 13 February, 2019
- Protein-Complex-Analysis, life-sciences: news-medical.net, Retrieved 3 August, 2019
- Challenges-of-Protein-Complex-Analysis, life-sciences: news-medical.net, Retrieved 30 March, 2019

## Chapter 6

Bioinformatics Software

There are a variety of different software which are used in the field of bioinformatics, such as EMBOSS, eProbalign and geWorkbench. The diverse applications of these bioinformatics software have been thoroughly discussed in this chapter.

EMBOSS

EMBOSS is a suite of programs developed to do molecular biology by EMBnet. EMBOSS is free software and although it is primarily developed for Linux it also runs in MS Windows and Mac OS X.

EMBOSS is comprised by hunders of programs and includes utilities to align sequences, translate RNA to proteins, identify protien motifs, analyze repetitions, calculate codon usage statistics and many more analyses.

Running EMBOSS Commands

The EMBOSS utilities have a Command Line Interface and all can be run from a text terminal. The documentation for the EMBOSS aplications is written taking into account mainly the Command Line Interface. But there are other ways to run the EMBOSS software.

There are web servers that provide the EMBOSS tools as a service. In those servers the EMBOSS utilities can be accessed as web pages and the server runs the terminal command and return the result.

Another interface to EMBOSS is Jemboss. Jemboss is a Graphical User Interface application, like any of the other common desktop applications.

Identify the Command

When we do not know which of the EMBOSS commands to use we can use another EMBOSS command, wossname, to search in the EMBOSS command list.

Get a list with all commands:

```
$ wossname
```

Look for command to do alignments:

```
$ wossname alignment
```

Getting help for a command:

The command manuals can be found in the web. They are also available in any EMBOSS installation by using the tfm command. For instance, to read the manual for the water command we could run:

```
$ tfm water
```

The manual reader is similar to the Unix less command. It is also possible to get a brief list of parameters available with the option "-help" and the complete list with "-help -verbose":

```
$ water -help -verbose
```

Example: change a sequence format.

To get familiar with the EMBOSS way of running the analyse we will change the format of a sequence file. EMBOSS has a program to modify sequence files: seqret seqret can change the sequence format, trim the sequences and reverse-complement them. Let's change the format of fasta sequence file to the GenBank format. If we run seqset in the command line it will ask for the input file and for the output file:

```
$ seqret
```

By default seqret won't change the sequence format unless we specify the output format that we want. To do it we have to write genbank::secuencia.gb. We can also tell seqret which is our input file:

```
$ seqret secuencia.fasta genbank::secuencia.gb
```

We could sent the result to standard output instead of a file. This is a general feature of all EMBOSS programs:

```
$ seqret secuencia.fasta genbank::stdout
```

To reverse and complement a sequence we can use the parameter -sreverse1:

```
$ seqret secuencia.fasta embl::stdout -sreverse1
```

Would the result sequence be different in this case? seqret can also trim the input sequence.

```
$ seqret secuencia.fasta embl::stdout -sbegin1 10 -send1 30
```

eProbalign

the eProbalign web server doubles as an online platform for post-alignment analysis. The heart-and-soul of the post-alignment functionality is the Probalign Alignment Viewer applet, which provides users a convenient means to manipulate the alignments by posterior probabilities. The viewer can also be used to produce graphical and text versions of the output.

Multiple sequence alignments are frequently employed for analyzing biomolecular sequences. Their application spans a wide range of problems such as phylogeny reconstruction, protein functional site detection, and protein and RNA structure prediction. The research literature is abundant with programs and benchmarks for multiple sequence alignment, particularly for protein data. Traditionally, ClustalW is the most popular program used for multiple sequence alignment; while BAliBASE is a likely the most commonly used benchmark of protein alignments.

MAFFT, Probcons and Probalign are recent alignment strategies that are among recent programs with the highest accuracies on BAliBASE and other common benchmarks (i.e. HOMSTRAD and OXBENCH. Both Probcons and Probalign compute maximal expected accuracy alignments using posterior probabilities. In Probcons, posterior probabilities are derived using an HMM whose parameters that have been estimated via supervised learning on BAliBASE unaligned sequences. Probalign, which is largely based on the Probcons scheme, derives the posterior probabilities from the input data by implicitly examining suboptimal (sum-of-pair) alignments using the partition function methodology for alignments. Probalign alignments have been shown to have a statistically significant improvement over Probcons, MAFFT and MUSCLE.

eProbalign also provides a convenient platform to visualize the alignment, generate images, and manipulate the output by average column posterior probabilities. The average column posterior probability can be considered a measure of column reliability where columns with higher scores are more likely to be correct and perhaps biologically informative.

Input Parameters

eProbalign takes as input unaligned protein or nucleic acid sequences in FASTA format. eProbalign checks the dataset to make sure it conforms with IUPAC nucleotide and amino acid one letter abbreviations. White space between residues/nucleotides in the sequences are stripped and the cleaned sequences are passed on to the queuing system. The user can specify gap open, gap extension, and thermodynamic temperature parameters on the eProbalign input page. The input page provides a brief description of the parameters (help link) and links to the standalone Probalign code with publication and datasets.

Figure: eProbalign input page.

The three Probalign parameters on the input page are used for computing the partition function dynamic programming matrices from which the posterior probabilities are derived. This is the same as computing a set of (suboptimal) pairwise alignments (for every pair of sequences in the input) and then estimating pairwise posterior probabilities by simple counting. The thermodynamic temperature controls the extent to which suboptimal alignments are considered. For example, all possible suboptimal alignments would be considered at infinite temperature, whereas only the single best would be used at a temperature of zero. The affine gap parameters are used for the pairwise alignments. Subsequently, Probalign computes the maximal expected accuracy alignment from the posterior probabilities in the same way that Probcons does.

Output and Alignment Column Reliability

The eProbalign output provides three options for viewing and analyzing the alignment. The alignment can be viewed in (i) FASTA text format, (ii) pdf graphical format, and (iii) the Probalign Alignment Viewer (PAV) applet. Each column of the alignment in the pdf file and in the applet is colored in a shade of red according to the average column posterior probability. Bright red indicates probability close to one whereas white indicates close to zero.

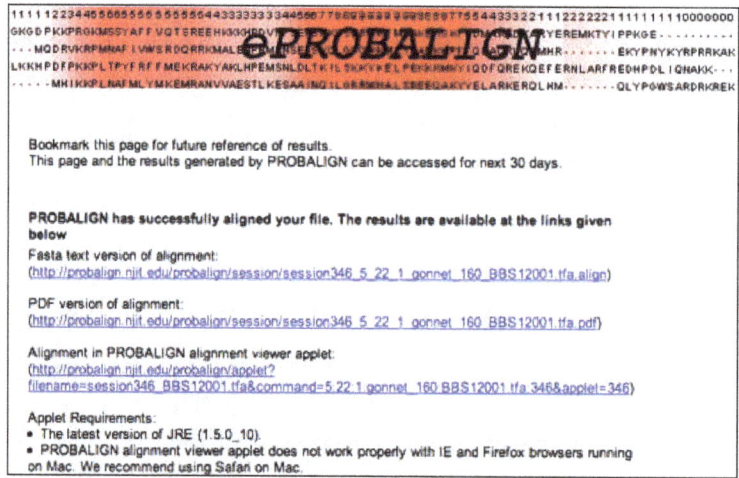

Figure: eProbalign output page indicating results are done.

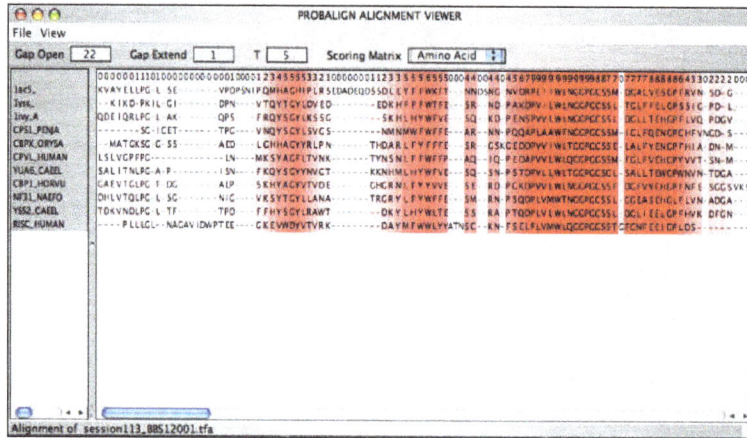

Figure: Probalign Alignment Viewer applet.

The average column posterior probability is defined as the sum of posterior probabilities of all pairwise residues in the column normalized by the number of comparisons. The top row of the alignment in the pdf and applet displays the average column posterior probabilities multiplied by ten and floored to the lower integer. For example, a score of 1 indicates that the probability is between 0.1 and 0.2.

The Probalign Alignment Viewer is a Java applet that provides basic manipulation of the alignment. Basic Java and browser requirements to use the applet are listed on the output page. With the applet the user can opt to view and save the alignment with column posterior probabilities above any specified threshold. This has the benefit of "cleaning up" the alignment by column posterior probabilities, which is unique to eProbalign. The applet also displays posterior probabilities of all columns in a separate window if desired and provides options to switch between the gapped and ungapped versions of the alignment.

Posterior Probabilities of Columns	
Column Number	Probability
1	0.084392
2	0.084724
3	0.083400
4	0.093505
5	0.089504
6	0.096759
7	0.129191
8	0.123973
9	0.124272
10	0.000000
11	0.114943
12	0.000000
13	0.079146
14	0.078184
15	0.000000
16	0.000000
17	0.000000
18	0.000000
19	0.000000
20	0.000000
21	0.000000
22	0.084132
23	0.099134
24	0.144801
25	0.000000
26	0.000000
Close	

Figure: Posterior probability of each column.

Server Implementation

We implement a first-in/first-out queuing system that receives requests for Probalign alignments and processes them accordingly. At most, eProbalign will run two Probalign jobs at once, and it will periodically check the queue for new requests. Alignments that take longer than some defined time limit are stopped and the user is advised to download and run the standalone version. This time limit will be increased as the server hardware is upgraded.

Scalability

Currently, eProbalign is installed on a dual processor 2.8GHz Intel Xeon machine with 2GB RAM. With these settings, eProbalign can usually align datasets of up to 20 sequences within one minute. Most BAliBASE 3.0 datasets from RV11 and RV12 also finish within one minute. We have also tested large datasets (in number and length of sequences) from BAliBASE RV30 and RV40 classes on eProbalign. BB30029 and BB30008 from RV30 contain 98 and 36 sequences with lengths from 431 to 852 and 400 to 1155 respectively, and BB40002 from RV40 contains 55 sequences with lengths ranging from 58 to 1502. When the server is idle, eProbalign finished in about 20 minutes on BB30008, 55 minutes on BB30029, and 30 minutes on BB40002. Results may take longer to finish when the server queue is full and multiple jobs are running simultaneously. However, the effect of parallel jobs will diminish as the server moves to a bigger machine in the near future.

.NET Bio

.NET Bio is a bioinformatics toolkit that was built using the Microsoft 4.0. NET Framework. It is designed for use by developers, researchers, and scientists, making it simpler to build applications to meet the needs of life scientists. This open-source platform features a library of commonly used bioinformatics functions plus applications built upon that framework, and can be extended by using any Microsoft .NET language, including C#, F#, Visual Basic. NET, and IronPython.

Users can perform a range of tasks with. NET Bio, including:

- Importing DNA, RNA, or protein sequences from files with a variety of standard data formats, including FASTA, FASTQ, GFF, GenBank, and BED.

- Constructing sequences from scratch.

- Manipulating sequences in various ways, such as adding or removing elements or generating a complement.

- Analyzing sequences by using algorithms such as Smith-Waterman and Needleman-Wunsch.

- Submitting sequence data to remote websites (for example, a Basic Local Alignment Search Tool [BLAST] website) for analysis.

- Outputting sequence data in any supported file format, regardless of the input format.

Like other frameworks (for example, BioJava and BioPython), .NET Bio can help reduce the level of effort that is required to implement bioinformatics applications through the provision of a range of pre-written functionality.

In addition to enhancements to the performance and capacity of the basic features contained in the previous version, the new version will provide a range of new features and demo applications. This includes:

- Access to advanced math functions by using Sho scripting

- A comparative DNA sequence assembler sample application

- A range of command-line utilities

.NET Bio is now in use by both academic and commercial organizations—including Microsoft—worldwide.

geWorkbench

geWorkbench (genomics Workbench) is an open source Java desktop application that provides access to an integrated suite of tools for the analysis and visualization of data from a wide range of genomics domains (gene expression, sequence, protein structure and systems biology). More than 70 distinct plug-in modules are currently available implementing both classical analyses (several variants of clustering, classification, homology detection, etc.) as well as state of the art algorithms for the reverse engineering of regulatory networks and for protein structure prediction, among many others. geWorkbench leverages standards-based middleware technologies to provide seamless access to remote data, annotation and computational servers, thus, enabling researchers with limited local resources to benefit from available public infrastructure.

A large number of bioinformatics techniques have been developed in recent years to serve the needs of biomedical research. The field is moving rapidly, with new and improved approaches appearing frequently. The fast pace of change and the technical sophistication of these approaches creates a barrier of adoption for ordinary biologists. The problem is exacerbated by the integrative nature of biomedical research, which requires combining data from multiple genomic/biomedical databases and using an array of advanced analyses, often available only in the form of command line programs. Additionally, due to their sheer size and dimensionality, analysis of genomic data sets can be computationally very demanding. It is unlikely that every biomedical researcher that would like to utilize such analyses will have access to local/ institutional hardware resources capable of supporting their execution. It is then important to facilitate sharing of public infrastructure through the use of technologies such as grid computing.

geWorkbench integrates many computational resources and makes them available through a unified user interface. For biomedical researchers with little or no computational training, this approach facilitates adoption by eliminating many steps that require programming skills (e.g. transformations from one file format to another; staging of databases and programs; dealing with operating system (OS) shells in order to execute command line programs). For software developers, geWorkbench provides an open source, component-based architecture that enables the addition of new functionality in the form of plug-in modules (which can further leverage existing tooling that streamlines access to server side components). Extensive documentation is available (manuals, online help and tutorials) to guide users in the proper use of the application. An innovative logging framework collects and mines data about how various modules are being utilized, offering the possibility for novice users to learn from their more advanced peers.

Design, Implementation and Support

geWorkbench comprises at present more than 70 distinct modules, supporting the integrated analysis and visualization of many types of genomic data. Users of geWorkbench can:

- Load data from disk or from remote data sources (such as the caArray microarray data repository at the National Cancer Institute, the PDB protein structure database; and sequence databases at the University of Santa Cruz and the European Molecular Biology Laboratory).

- Visualize gene expression, molecular interaction networks, protein sequence and protein structure data in a variety of ways.

- Access client- and server-side computational analysis tools such as t-test analysis, hierarchical clustering, self-organizing maps, analysis of variance (ANOVA), regulatory and signaling network reconstruction, basic local alignment search tool (BLAST) searches, pattern/motif discovery, etc.

- Validate computational hypothesis through the integration of gene and pathway annotation information from curated sources as well as through enrichment analyses.

Several modules have been developed in collaboration with investigators from the Center for the Multi-scale Analysis of Genomic and Cellular Networks, one of seven National Centers for Biomedical Computing; The mission of the MAGNet Center is to provide the research community with novel, structural and systems biology methods and tools for the dissection of molecular interactions in the cell and for the interaction-based elucidation of cellular phenotypes. Other geWorkbench modules are wrapped versions of pre-existing third party software tools such as Cytoscape, Ontologizer and GenePattern.

Table: geWorkbench plug in modules based on MAGNet tools

Plugin Name	Description
ARACNe	Prediction of transcriptional interactions from gene expression data.
MINDy	Identification of modulators of transcriptional regulation.
Matrix REDUCE	Physics-based prediction of DNA-binding sites.
MEDUSA	Machine learning-based prediction of DNA-binding sites.
MarkUs	Functional annotation of protein structures.
Pudge	Protein structure prediction from sequence
SkyBase	Database of predicted protein structure models.

Software Architecture

geWorkbench is designed around a component-based architecture whose main purpose is to facilitate the expeditious integration of algorithms and data sources as plug in modules.

Server Side Computing

The computational, administration and storage requirements of many bioinformatics resources developed by MAGNet (such as large genomic databases and CPU/memory-intensive algorithms) make deployment on a desktop computer infeasible. To make them available to geWorkbench (as well as other clients), such resources are deployed as programmatically accessible grid services using caGrid, the grid middleware layer of the caBIG® initiative. An attractive feature of caGrid is that if offers tooling, which streamlines the process of defining, deploying and registering caGrid-aware services. caGrid services have been developed by other participants of the caBIG® program as well and geWorkbench provides access to many of them (caArray, the Cancer Gene Index, the NCI Pathway Interaction Database, etc.)

Documentation and user Support

geWorkbench is accompanied by detailed end-user documentation, demonstrating how to perform common analysis tasks and explaining the theoretical underpinnings of many analysis modules. Community support is provided by the caBIG® Molecular Analysis Tools Knowledge Center.

Permissions

All chapters in this book are published with permission under the Creative Commons Attribution Share Alike License or equivalent. Every chapter published in this book has been scrutinized by our experts. Their significance has been extensively debated. The topics covered herein carry significant information for a comprehensive understanding. They may even be implemented as practical applications or may be referred to as a beginning point for further studies.

We would like to thank the editorial team for lending their expertise to make the book truly unique. They have played a crucial role in the development of this book. Without their invaluable contributions this book wouldn't have been possible. They have made vital efforts to compile up to date information on the varied aspects of this subject to make this book a valuable addition to the collection of many professionals and students.

This book was conceptualized with the vision of imparting up-to-date and integrated information in this field. To ensure the same, a matchless editorial board was set up. Every individual on the board went through rigorous rounds of assessment to prove their worth. After which they invested a large part of their time researching and compiling the most relevant data for our readers.

The editorial board has been involved in producing this book since its inception. They have spent rigorous hours researching and exploring the diverse topics which have resulted in the successful publishing of this book. They have passed on their knowledge of decades through this book. To expedite this challenging task, the publisher supported the team at every step. A small team of assistant editors was also appointed to further simplify the editing procedure and attain best results for the readers.

Apart from the editorial board, the designing team has also invested a significant amount of their time in understanding the subject and creating the most relevant covers. They scrutinized every image to scout for the most suitable representation of the subject and create an appropriate cover for the book.

The publishing team has been an ardent support to the editorial, designing and production team. Their endless efforts to recruit the best for this project, has resulted in the accomplishment of this book. They are a veteran in the field of academics and their pool of knowledge is as vast as their experience in printing. Their expertise and guidance has proved useful at every step. Their uncompromising quality standards have made this book an exceptional effort. Their encouragement from time to time has been an inspiration for everyone.

The publisher and the editorial board hope that this book will prove to be a valuable piece of knowledge for students, practitioners and scholars across the globe.

Index

P
Pathway Analysis, 122, 174
Pedigree Analysis, 137
Phage, 2
Pharmacogenomics, 124, 127-128
Principal Component Analysis, 55, 82
Protein Expression, 175-176
Proteomics, 1, 3, 5, 132, 137, 139, 202, 208, 210

R
Recombination, 95, 97, 150
Recurrent Neural Networks, 89

S
Sequence Similarity, 2, 13, 145, 151-153, 155-156, 205

Sequencing Library, 186, 191
Serial Analysis Of Gene Expression, 3, 177, 183, 185
Single Nucleotide Polymorphisms, 179
Smith-waterman Algorithm, 26, 28-29, 98-99, 144
Structural Bioinformatics, 118, 133-134

T
Tandem Repeat, 131
Transcription, 51-53, 78, 120-121, 132, 165-166, 172, 175-176, 184-186, 189, 197
Transcriptomics, 1, 132, 180, 185-186, 191, 198
Translational Bioinformatics, 118, 122-124

W
Whole-genome Sequencing, 127, 178

www.ingramcontent.com/pod-product-compliance
Lightning Source LLC
Chambersburg PA
CBHW082047190326
41458CB00010B/3478